Curriculum Dynamics

COUNTERPOINTS

Studies in the
Postmodern Theory of Education

Joe L. Kincheloe and Shirley R. Steinberg
General Editors

Vol. 200

PETER LANG
New York • Washington, D.C./Baltimore • Bern
Frankfurt am Main • Berlin • Brussels • Vienna • Oxford

M. Jayne Fleener

Curriculum Dynamics

Recreating Heart

PETER LANG
New York • Washington, D.C./Baltimore • Bern
Frankfurt am Main • Berlin • Brussels • Vienna • Oxford

Library of Congress Cataloging-in-Publication Data

Fleener, M. Jayne.
Curriculum dynamics: recreating heart / M. Jayne Fleener.
p. cm. — (Counterpoints; vol. 200)
Includes bibliographical references and index.
1. Curriculum planning—Social aspects. 2. Curriculum change—Social aspects. 3. Education—Philosophy. 4. Postmodernism and education.
I. Title. II. Counterpoints (New York, N.Y.); vol. 200.
LB2806.15 .F57 375'.001—dc21 2001038809
ISBN 0-8204-5540-7
ISSN 1058-1634

Die Deutsche Bibliothek-CIP-Einheitsaufnahme

Fleener, M. Jayne:
Curriculum dynamics: recreating heart / M. Jayne Fleener.
–New York; Washington, D.C./Baltimore; Bern;
Frankfurt am Main; Berlin; Brussels; Vienna; Oxford: Lang.
(Counterpoints; Vol. 200)
ISBN 0-8204-5540-7

Cover art by Eric Schol
Cover design by Joni Holst
Cover concept by M. Jayne Fleener

Table of Contents

Tables and Figures

Foreword
By
William Doll

"America is continually struggling for its soul"
(Gunnar Myrdal, *An American Dilemma*, 1962/1944, p. 4)

Jayne Fleener feels this struggle deeply—in the classes she teaches, in the books and articles she writes, and in the life she lives. With her background in philosophy, Jayne recognizes only too well that we who live in Western nations have had our thoughts caught in the vise grip of rationalistic logic, a logic she says that is characterized by "domination" (Chapter Two). Today we—students, educators, parents—are dominated by tests and standards. They, not we, determine our worth, our being. They are heartless. We have given up our hearts and souls to the "gods of measurement." But Jayne is not bereft by the trauma of this current situation. Her strong interest in mathematics, especially the mathematics of turbulence, has led her to see that it is relationships not brute facts that are the "really real," to use Whitehead's felicitous phrase. Thus Jayne encourages us to embrace our contemporary challenges and difficulties with a faith that what we now see as chaos really contains, if we look closely enough, a "hidden order": a pulsating, heartfelt, dynamic, soulful, and transformative order.

If we are willing to exchange—and Jayne believes we must—our modernist lenses with their binding frames and limited perspectives for postmodernist ones, then we shall see anew. We shall see new logics, ones based on a nesting of relationships wherein holistic and organic systems, not isolated elements, become prominent. With such a vision in place, Jayne's last chapter draws on Starhawk (1988) who says:

> I am aware of the world and everything in it as alive, dynamic, interdependent, interacting, and infused with moving energies: a living being, a weaving dance. (p.11; in Fleener, p. 192)

It is this attitude that reflects Jayne's outlook on life and hope for our schools—schools she would like us to consider as "learning organizations," not institutions of domination and control.

Jayne has divided her book into three sections: one giving us a perspective on where we are at present and the role modernist thinking and its logic of domination have played in getting us where we are. Jayne also looks to the future and how it is being shaped by post quantum physics and new directions in evolutionary biology. Section two focuses on postmodern logics, systems, learning, and meaning. In her third section Jayne focuses on creation, curriculum, and change with the plea for us to (re)create heart in our process of developing curricula, schoolings, strategies for instruction (and living). In each section Jayne makes provocative points. In the first section, entitled *Erwartungshorizonten* for "horizons of expectation" (p. 12), Jayne well shows us that our horizons for reform are limited by our modernist language and logic, both heavily influenced by classical Newtonian science and Cartesian individualized rationality. Our modernist language honors the mechanical and closed (p. 15), just as our modernist logic, caught in an oppositional frame of either/ors and right/wrongs, honors its own dichotomy of "domination/subordination" (p. 43). School reform that encourages domination/subordination and assumes a closed and mechanical reductionism is to be feared more for its success than for its failure.

Bleak as the picture of schooling is—and has been for centuries back to Ramus, Bacon, Comenius, Descartes, Locke, all with their emphases on schooling being simple memorization, mechanical repetition, and a transmitive pouring or "depositing" of knowledge into blank or empty minds—Jayne is not dispirited. She has a vision of curriculum with heart, a dynamic curriculum that is pulsating with excitement and creativity. Such a curriculum is not one that has more or different "stuff" in it; it is a curriculum with a very different "attitude" (p. 192) or frame. Such a curriculum is based on the dynamics of life; on process thought; on open, complex systems; on meaning-making. Such a curriculum does have an attitude, an attitude toward life and living that opens up new and as yet unthought of possibilities. To see these potential possibilities we need to replace our modernist lenses and logics with postmodernist ones.

In her second section, "Technique—Holographic Images," Jayne explores logics that do not lead to a domination/subordination split. These are the postmodern logics of relationship, process, dynamic systems, and meaning. What these logics have in common is a sense of holographic holism; that is, they see life, learning, schooling, and reality as seamless, but having different interrelated, interactive, integrated perspectives. Of all the ways to describe this new approach to life and learning, the one that resonates strongest with Jayne and which has provided a new attitude for her in working with her students is that of relationships:

> Thinking of students, learners, and schooling as *relationships and contextual*...[not as "things"] has completely affected what I feel is important in my classroom, how I...approach instruction, and how I view assessment. (p. 78; emphasis in original)

Dewey's lifelong objection to the logical splits of mind/matter, either/or, right/wrong, teacher/student, child/curriculum, knower/known were that these dichotomies (1) privileged one set of any pair over the other, and (2) limited participatory interaction between the two. Instead Dewey proposed a logic, not of domination/subordination but of holistic relationships, inquiries, transformations. It is this switch of attitude that has meant so much to Jayne and her ways of dealing with her students—seeing them and herself as humans engaged in the common quest of learning and living. Many have written of this transformative, experiential frame of Dewey's, but none I know of have dramatized it quite as Jayne has in quoting from Karen Warren's (1996) description of climbing a mountain. Jayne says that at first Karen "set out to conquer the mountain" with much use of the supporting rope and the acquisition of many bruises (p. 92 in Fleener). The next day she chose to remove the rope and climb alone, just herself and the mountain, in a symbiotic relationship:

> I closed my eyes and began to feel the rock.... I began to talk to the rock...as if the rock were my friend.... Gone was the determination to conquer.... I wanted simply to work respectfully with the rock as I climbed. (p. 93, in Fleener; pp. 26-27 in Warren)

It is this movement from a logic of domination, quite Western and male, to one of cooperation and transformation that so inspires Jayne and gives her a new outlook on the possibilities inherent in teaching, education, schooling, and curriculum.

The foundations for this vision lie, Jayne believes, in A. N. Whitehead's process philosophy and in John Dewey's logic as inquiry, as well as in the "new sciences" of chaos and complexity, the logic of systems, and Ludwig Wittgenstein's language games and forms of life. In concert, these activities and their techniques form a sort of new Grand Narrative: a narrative with heart but without a set center, or as Jayne likes to phrase it, without an "essential 'secret in the middle'" (p. 136). There is no middle, there is no secret.

Western philosophy from Plato on has assumed that some underlying truth centers our daily lives, a truth we, the reasonable, should be able to (but never do) find. Jayne quotes Dewey on this point:

> The old center was mind [complete in itself] knowing...an attendant external material equally complete in itself. *The new center is indefinite interaction taking place within a course of nature which is not complete.* (p. 87 in Fleener; pp. 290-292 in Dewey's 1929, *Quest for Certainty*, emphasis by Fleener)

Thus, the old worldview, its *Weltbild*, premised a set, uniform, hidden, regulatory order—a collection of "underlying laws of nature that are knowable" (p. 126). The new *Weltbild*, Grand Narrative if you will, that Jayne sees, rejects this "secret in the middle" perspective. Rather, the new perspective is one of seeing all in local context (Wittgenstein), in relations (Whitehead), in the practicalities of life (Dewey and Wittgenstein), in the dynamic flux of emerging and creative systems (chaos and complexity). Jayne believes that this new way of "seeing"—interrelating local contexts rather than finding set universals—will provide us a framework, a *Weltbild*, for doing schooling in a new manner. However one feels about this vision, and it is one I favor strongly, one has to be impressed with how well and clearly Jayne explains Dewey's concept of logic, Wittgenstein's forms of life, autopoetic systems, and the intricacies of chaos and complexity theories. No one I know has done this better.

As a curriculum theorist, mathematics educator, teacher, and learner—one who learns from her students—Jayne is much interested in and committed to the process of schooling. This is a process she believes should be creative, imaginative, and soulful—not deadening—and to this process she devotes her last three chapters. In some ways it is absurd to consider schools as anything but learning organizations—this is their function. But, as Jayne points out, this is not their "spirit," nor their "attitude," to return to the Starhawk quotation which I use to begin this Foreword and which Jayne uses to end her book. Schools are places, cement and concrete places, where bits of knowledge are dispersed via a structure called curriculum. Both schools as places and curricula as lists of knowledge disbursements accept, indeed honor, the root metaphor of "things"—machine-oriented things, so common to modernist thinking. Schools as *thing-places* dispense bits of knowledge as *thing-items*.

Jayne's root metaphor—a living, beating, dynamic heart, alive with life and its never-ending possibilities—considers both organization and learning in different terms. Organization for Jayne is not a "thing" with its engineering frame, mechanical, cause-effect way of operation, and rule bound sense of mission. Rather, organization for Jayne lies with Whitehead's sense of process, Dewey's sense of situation and experimentation, Wittgenstein's sense of forms of life, system theory's sense of adaptation, self-organization, and creative transformation. In all these senses there is a metasense of a dynamic form: bounded but infinite in its possibilities (as with the Mandlebrot set), breathing with heart and soul but without a center (domination and control are dispersed not centered), and maintaining its continually evolving coherence through stochastic change. This is life in all its throbbing complexity.

Learning for Jayne is neither the accumulation of dispersed facts nor the acquisition of those ideas Whitehead (1929) labels as "dead, inert" (*Aims of Education*, Ch. 1). Rather, learning for Jayne is deep and meaningful, the sort that comes with students and teachers "seeing" and exploring the complexities of any situation or event through its interrelationship with other situations or events. Learning is thus holographic and the techniques for acquiring this sort of learning are those of seeing synoptically, of doing reflectively, and of living soulfully. Here is a challenge Jayne gives to all of us who are educators, curricularists, teachers, indeed livers of life.

Obviously it is my honor to write a Foreword to this marvelous and inspiring book. I'd like to close with one, most personal, comment. What Jayne has done with my concept of a good curriculum—its Richness, Recursion, Relations, Rigor (4Rs)—is a tribute to her ability to take a simple idea and create from it more than was originally there. She is truly a dynamic curricularist whom I am honored to call friend and colleague.

Acknowledgments

The intellectual, spiritual, emotional, and social support that needs to be acknowledged, especially for a first book, is tremendous. Rather than tracing back all who have had an impact on my ways of thinking over the past several decades, however, I will focus my comments of gratitude to those who have had a direct influence on this work. To all the unnamed others, however, I acknowledge and appreciate your contributions. Without friends, family, colleagues, and other searchers, I would not be the person that I am.

To my many colleagues in different professional organizations, I owe a debt of gratitude for your support, your attendance at my presentations, and your comments, questions, and nurturing challenges as I developed many of these ideas. Especially to my colleagues in the Research Council for Mathematics Learning and the Chaos and Complexity Theories Special Interest Group of the American Educational Research Association, I want to express my appreciation for your tolerance and encouragement as I tried out many of these ideas on you. To Bill Doll in particular, a special kind of mentor and friend: I have benefited tremendously from our discussions and the support you have offered me over the last few years.

To my colleagues at the University of Oklahoma (OU) I owe so much. To Anne Reynolds, for being an outstanding colleague, intellectual partner, and friend I owe a lot. Our work together in developing a program and working with our graduate students has helped me envision and challenge how curriculum may be. To Neil Houser, Susie Laird, Pamela Fry, and Deborah Rodgers I would like to say how much I have appreciated our discussions, your willingness to share ideas, resources, books, articles, and your support of my efforts in every aspect of my job. We have written together, shared manuscripts, and explored ideas all of which have been so important for my own growth and explorations. For whatever confusions or inaccuracies I still convey in this book, I apologize.

To others in the department of Instructional Leadership and Academic Curriculum, I have appreciated the leadership opportunities and support you have shown me over the years. And to my dean, Joan K. Smith, I have learned from you what it means to be a leader in a learning organization, to value faculty governance, and to work with faculty as individuals. I have been fortunate to have been supported, encouraged, and valued by my colleagues at OU and to those named and not named I offer my sincere appreciation for all you have taught me and all you have meant to me.

Special recognition goes to those for whom this book is really written. I have been fortunate over the years to work with many outstanding graduate students and companions in our mutual searching and explorations. Many of the ideas presented in this book have been tried out in a variety of the graduate level classes I have taught including the chaos, theoretical issues, technology, and curriculum classes. To those who have asked questions, challenged assumptions, and demanded practicality, I owe much. To the students who have directly contributed to the completion of this book, my best expression of gratitude is my hope that you have learned as much as I in the process. To Sue Erwin, John Erwin, Dana Cesar, Megan Che, Mary Reeves, Deborah Shinn, Janice Berry, and Valerie Rogers, your contributions have been tremendous and I will forever be grateful for your thoughtful, caring, and kind comments during the revising process. While any errors or confusions are completely my responsibility, you have done much to make this a clearer, more inclusive and readable book. To Stacy Reeder, Keith Adolphson, and Darlinda Cassel, your work both on this book and in working as graduate teaching assistants and research assistants in mathematics education, I owe a special acknowledgment of gratitude. You have taught our classes, read my papers, and explored these ideas with me. But you have also gone the extra mile to think of ways to improve the program and contribute your time and energy to our common efforts.

To Melissa Smith, Julie Comer, and Joann Nilsson, I want to express my appreciation for all you do to keep the office and my life running smoothly. Especially as I made final revisions, printed numerous extra copies, and rewrote and needed even more copies, your help allowed me to keep focused.

To my first teacher in living and loving, I cannot express the depth of gratitude I owe and love I have for my mother, Marion. She has always been a model of the kind of person I want to be. Her unconditional love, support, appreciation, and respect for me have given me the courage and confidence to negotiate the world. Equally important, however, is what she has taught me about caring and loving others. Her aura of love and her connected way of being are models for recreating heart.

To all of my parents I owe so much. Having a "modern" family, I have been fortunate to have had my own two sets of parents as well as my in-laws who have supported, appreciated, loved, and modeled lives of learning, caring, and giving. To John, Janet, Larry, Bob, and Wynell, thank you for always being there and for accepting me as I am.

And, of course, to my husband Mike and son James, I owe a great deal and also offer my apologies. You have most directly had to sacrifice your weekends, evenings, dinners, and family time as I completed this book and juggled all of my other professional responsibilities. But you didn't complain. You cherished, nurtured, and loved me throughout the process. Because of your sacrifices and unconditional love and support, I dedicate this book to you.

Finally, I would also like to thank Eric Schol for creating the fractal used on the front cover. The Mandelbrot set has special significance as a complex phenomenon. He specifically revised his image for the book at my request. He reserves all rights for this image. His fractal images can be seen at: www.snt.utwente.nl/~schol/gallery/.

Introduction

Ever since I began teaching, I have asked the same series of questions. What did I want the students to know and to be able to do when they completed my class? What did they want and feel they needed to know? What was the purpose of the class? How did the class fit with my students' other educational experiences and goals? And ultimately, the BIG questions kept coming up: Education for what purpose? Schooling for whom? Why?

I have been fortunate to have always taught in environments where I could ask these questions and derive my own answers. Many teachers are not so fortunate. Part of how I approached my own answers to these questions was a result of the classes I was teaching. I taught mathematics and logic very differently than I taught computer science, at least at first. It was a revelation to me when a student who had me in high school geometry as well as AP Computer Science noted how differently I taught the two classes. Why couldn't I teach mathematics more like I taught computer programming? In the latter class, I gave students interesting problems and, based on their approaches to the problems, they learned various programming techniques and structures. I never expected students to come up with the same solutions or even to use the same data or control structures in the programs they constructed. It didn't matter whether some students who found the problem particularly appealing spent hours upon hours making modifications to their programs and adding complexity while others merely solved the given problem as stated. Why couldn't mathematics teaching and learning be more like that?

A second epiphany occurred while I was exploring topics in chaos mathematics and the complexity sciences. While reading James Gleick's classic book (1984) *Chaos: The Making of a New Science*, and Margaret Wheatley's book (1992) *Leadership and the New Science,* I realized I had been focusing on the wrong things in my teaching. From the complexity sciences, especially quantum physics, I learned the importance of shifting perspectives from looking at "things" to understanding and developing "relationships." Rather than considering *what* mathematics was being learned (or how to measure learning), I shifted my attention to looking at relationships. What were my students' relationships with mathematics, with me, and with each other? This simple change in perspective opened up an entirely new door and recalled my earlier interest in Whitehead and process philosophy.

I continued to explore the complexity sciences literature as well as the educational literature where I found these same or very related ideas. As I envisioned what relational teaching might be like, I revisited Dewey and Whitehead and became better acquainted with feminist theories, poststructuralism, postmodernism, and phenomenology. I also recalled and expanded my readings in philosophy. Process versus analytic philosophies, problems in the philosophical foundations of mathematics, philosophy of science, and even the debates about the death of philosophy became interconnected with my search for a relational approach to education and the curriculum. Friendships with other seekers, Bill Doll, Jim Garrison, Susie Laird, John St. Julien, among many, pushed me to explore my intuitions and look in the dark corners of my philosophical past.

This book approaches the curriculum field from a wide range of academic and content area backgrounds. It is also intended to advance discussions about and provide background in the philosophical and historical foundations of education. Finally, I hope those outside of education engaged in complex adaptive systems (CAS) research will find the theoretical and philosophical discussions of CAS both enlightening and foundational for future directions and approaches their investigations may take.

The title of this book, *Curriculum Dynamics,* is intended to capture the process and complexity of curriculum work as it relates to curriculum, the child, and society. It also is intended to include complex adaptive systems and transformational perspectives of curriculum as a social phenomenon and meaning system. Integral to a dynamics perspective is that system features cannot be studied piecemeal. Understanding the complexity of the curriculum requires "taking a step back" and exploring how systems qua systems operate. No meaningful understanding of schools or the school's curriculum can result from studying piecemeal the curriculum, teacher preparation, teacher professional development, management structures, and decision-making models in isolation. Understanding schooling and the curriculum from systems, process, and meaning perspectives requires adopting dynamical approaches. Curriculum dynamics is thus a way of exploring the complexity of schooling as a social institution and curriculum as self-organizing and emergent. Curriculum will be seen to be the heart of schooling and curriculum dynamics the breath of life of the social and organizational matrix that includes schools.

As Dewey describes in *Art as Experience* (1934), art serves two functions: as a reflection and outlet of the artist's experience and as the impetus for the experience of others. The work of art is not the product, the paint on the canvas, but the composite and multiplicity of experiences associated with it. The subtitle of this book captures the various voices of Dewey, CAS research, process philosophy, and postmodern ideas about relationship and meaning. Our understanding of the complexity of the curriculum entails a holistic, multiperspectival, holographic approach to exploring our own experiences with schooling. Revisioning schools from historical, relations, systems, and meaning perspectives will entail experiencing schools in different ways.

"Recreating heart" can be read in a variety of ways. Recreating heart is my vision of putting the heart back into schooling. Gandhi's notion of *Satygagraha* captures part of this meaning. *Satygagraha* is a plan of action for personal and social transformation. It connects meaning, truth, and spirit by emphasizing the "soul" work of transforming society. We have lost a sense of our purpose, our essence of schooling, the "soul" work of teaching and learning in dynamic relationship with society and the individual. The "heart" is the life-force of our being. Whether or not the heart is truly the origin of our feelings of love, pain, and emotion, it is the metaphor we use most often. When love is shattered, we are said to have a broken heart. When we care deeply about an issue or are moved by something, our heart is said to swell up with pride or emotion. Recreating heart is a reinvention of the passion for and love of learning that seems to be missing in our standards-driven curriculum. Recreating heart is the *Satygagraha* of transforming schooling. It is the reasserting of meaning, purpose, value, and care as the driving forces of the curriculum. It is the integration and holistic understanding that teaching, learning, society, school, and the curriculum are different aspects of one basic meaning structure.

The subtitle of the book can also be read as "Re-creating He-Art" as a critique and rejection of the male-oriented logic of domination inherent in modern thinking. The "he-art" of the curriculum is built upon hegemonic logical structures that are oppressive to women and non-Western culture. Similarly, "Creating He-art" is a third reading of the title meant to capture that, while there are elements of the male-dominated perspectives which are incorporated into and transcended in my thinking, and while we are recreating the art as described above, I also feel the book is creating something new, exploring ideas about process, relations, systems, and meaning as underlying logics of a postmodern perspective. Mead argues in *The Philosophy of the Present* (1932) that all we have is the present. Any present contains a past which itself varies as the present itself varies and with each new present, new pasts will arise. The present "marks out and in a sense selects what has made its peculiarity possible" (p. 23). The he-art being created maintains the gender identification only to emphasize the shifting power relations and meaning structures, traditionally, at least in the West, focused on malestream thinking, inherent in postmodern, process thought. Creating he-art indicates the transformation of society from traditional malestream thinking to process meaning systems and adaptive cooperation. Finally and ultimately, this book is about creating heart, that is, planting the seeds for others to engage in and grow curriculum futures that transcend our modernist origins and reconnect with meaning, purpose, and value.

The conundrum of a curriculum book that purports to emerge from postmodern perspectives is that, while change oriented, accounting for change requires non-essentialist or non-reductionistic ways of understanding and ultimately directing change. One of Dewey's ghosts (see Doll, 2002) is precisely the fact that his vision for schooling, now over a century old, has yet to be realized. How can we re-frame our ideas about the curriculum from a postmodern perspective? How can we talk about meaning and purpose in education if these very ideas must evolve for each individual? How do we discuss the basis for a postmodern curriculum without relying on first principles, established goals, or standards?

The book is set out in three sections. I am not an artist and apologize to those whose expertise would find my visioning of the artistic process naive, but the titles of the three sections of my book, "perspective," "technique," and "creation," are intended to capture the processes involved in creating art. Perspective defines or directs our queries. It also reflects the times in which we live. For this reason, the first section of the book will include an exploration of the modernist mind-set and the inherent logic of domination underlying schooling and the curriculum.

Underlying this historical approach is *Erwartungshorizonten,* or horizons of expectation. *Erwartungshorizonten* allows us to theorize about the past while orienting us, at the same time, to the creation of possible curriculum futures. The first three chapters of the book therefore provide the perspective from which our curriculum futures may evolve. Scientific, social, and philosophical revolutions leading to process philosophies, pragmatism, and scientific revolutions will be explored as important seeds to the *Erwartungshorizonten* perspective and developing our postmodern perception.

The second section of the book focuses on the technical aspects of a postmodern curriculum grounded in relational, systems, and meaning logics. Aristotle used the term *techne* in association with the practice of an art. While *techne* is a skill, the development of *techne* requires *poietike* or creation.

> This form of action (*poietike*), which is dependent upon the exercising of skill (*techne*), always results from the idea, image or pattern of what the artisan wants to make.... When action is informed by a technical interest (that is, the disposition of *techne*), therefore, it is constituted by a number of elements. These are the *eidos* (the guiding idea) and the *techne* (the guiding disposition) which together provide the basis for *poietike* ("making" action). (Grundy, 1987, pp. 22–23)

The four chapters comprising the *techne* of my curriculum approach, Chapters Four through Seven, build upon the perspectives of relational logic, the logic of systems, and the logic of meaning. These approaches, in direct contrast to the logic of domination and the system of philosophy emerging from the Cartesian method, provide multiple perspectives of postmodern ideas about social systems. The logics of relationship, systems, and meaning together comprise a postmodern condition that expresses our own current uncertainties and difficulties with the conflicts and contradictions associated with our modernist roots. Adopting these underlying ways of valuing and directing inquiry has implications for our relationships with each other, our physical world, and our world of ideas and may serve to help us transcend our modernist origins.

Chapter Four explores the logic of relationship, which emerges from Whitehead's process approach and Dewey's approach to inquiry. The perspectives of Dewey and Whitehead serve as the basis of a relational logic that fundamentally challenges modernism's logic of domination. The interplay among religion, education, spirituality, and science becomes more holistic from the perspective of relational logic.

Chapter Five offers a perspective of the curriculum that reflects the relational logic described in Chapter Four and develops a systems logic utilizing approaches from the complexity and chaos sciences. Systems logic provides images and metaphors that significantly change how we look at schooling and the complex relationships among the child, curriculum, and society.

Chapter Six offers a third voice necessary for a vision of a postmodern curriculum. Postmodern language theory, emerging from Wittgenstein's language-games, offers a third technique for understanding our postmodern curriculum. The dynamic relationships among the logics of meaning, relationship, and systems will be explored as constituents of a postmodern perspective.

Chapter Seven extends the logics of relationship, systems, and meaning to social systems as learning organizations. Schools as learning organizations, capable of adaptive, self-organizing, collaborative, autonomous interactions, more adequately reflect the postmodern perspective and recreate the heart of learning within the school context.

The third part of the book, the *poietike*, requires using the *techne* of relationship, systems, and meaning perspectives to explore a postmodern curriculum. The *eidos* or guiding idea of this approach has implications but not prescriptions for action. Chapter Eight will continue this theme to consider how schools can become learning organizations capable of creative self-organization and emergence. The role of the curriculum for schools as learning organizations will be described. Doll's (1993) curriculum matrix will be explored as the *eidos* for a postmodern curriculum with self-organizing capabilities. The final chapter of the book will reconnect with *poiesis*, that is, will extend the discussion from Chapter Eight to address how the curriculum can become the guiding force of transformative change in schooling. While the conundrum of postmodernism makes prescriptive solutions to problems impossible, the "soul" of schooling can be identified as the curriculum. Transforming schools can be approached from a postmodern perspective with underlying logics of relationship, systems, and meaning. The importance of curriculum work to transform schooling has not been recognized nor strategized to the extent explored in this book.

My hope for those who read this book is that they find meaning, purpose, and value, not answers, prescriptions, or formulas to follow. For all who share my vision and have hope for schools with heart, my desire is that the postmodern logics of relationship, systems, and meaning will allow them to change their way of seeing necessary for altering ideas about schooling. Ultimately, recreating heart entails exploring meaning, purpose, and value, understanding and challenging our modernist assumptions, and creating a new way of seeing schooling, including teaching, learning, society, and the curriculum.

How do I achieve this goal? The only thing I can offer is a vision. The vision I offer is the picture that makes sense to me. I in no way purport to have THE answer or THE perspective. I also do not wish for followers or devotees of these ideas. I hope in sharing my journey and experiences that I will have more companions in and continued dialogue on the search for and reinventing of the heart of education.

PART I: PERSPECTIVE

Erwartungshorizonten

CHAPTER ONE
Origins and Cornerstones of Modernity

Anyone who has spent time with a preschooler can recall with bemusement (or irritation) how often conversation was dominated by an unending litany of "why's." However, only when coaxing our own imbedded truth from all the things that surround us is the correctness of the child's earliest intellectual forays most evident. We can never afford to abandon the profundity of "why."

This book is about envisioning a curriculum for postmodern (and post-postmodern) times. Much has happened in the 20th century. We marvel at how the world has changed during the lifetimes of our parents and grandparents. Helping our own children in their search for meaning and security is often a challenge for us as we ourselves struggle to keep up with rapid technological changes, social diversity, and volatile markets. Growing up during the cold war, we baby boomers still carry the perverse idea that there must be a "bad guy." Is that how we want our children to grow up, looking for the boogie man in closets, under the bed, or in the church down the street? What we expect from our answers reveals an underlying belief about our place in the cosmos. While the search for certainty, security, and order may prompt the questioning of our youth, our more mature selves may appreciate that the search for meaning and understanding is one of creation rather than discovery. Meaning and language are not anchored in some universal actuality but are themselves avenues to abstraction.

Following a Wittgensteinian path, meaning is conveyed and transformed through language and our creative invention of ideas. Definitions of words can only be understood in the history and context of their use rather than an appeal to some underlying meaning or fixed reality. Meanings evolve as use and context change. Exploring the possibilities of and creating curriculum futures, we must address our own boogie man; those ideas, practices, and goals that have constrained our ability to change, adapt or create a new reality for schooling. Our boogie man has perhaps been our own fears of change and our own unquestioning acceptance of the times in which we live.

Epochs as historical periods of time also can only be understood from the perspective of meaning. While often identified with great accomplishments of inventions, constructions, or social organizations, what distinguishes one epoch from another is the underlying and driving search for meaning. To understand our modernist origins in order to evolve our curriculum futures requires a reconstruction of our curriculum pasts and an understanding of modernism as the prevailing system of meaning.

The delineation of "modernism," according to Toulmin (1992), requires a perspective of what he calls *Erwartungshorizonten* or "horizons of expectation." *Erwartungshorizonten* is grounded within an historical perspective, as described by Toulmin (1992):

> To form reasonable and realistic "horizons of expectation" today, we must therefore begin by reconstructing an account of the circumstances in which the modern project was conceived, the philosophical, scientific, social, and historical assumptions on which it rested, and the subsequent sequence of episodes that has led to our present quandary. (p. 3)

Erwartungshorizonten both contextualizes and limits our vision but also emphasizes future action. Whitehead's process approach (1925) similarly emphasizes the dialectic between past and future orientations:

> There are two ways of reading history, forwards and backwards. In the history of thought, we require both methods. (p. 3)

I will address the origins and underlying meaning structures of modernity and ideas about the modern curriculum not merely for the sake of historical completeness but as the basis for our own dialectic with the future, our *Erwartungshorizonten* of curriculum futures evolving from curriculum present and past. The approach of *Erwartungshorizonten* is a reconstruction of the present curriculum circumstances from the perspective of understanding the past and with a vision toward the future. Attempts to capture the meaning systems and ideas of the past, however, are in themselves derivations of meaning.

There is no essence, no single truth to be captured when we try to make sense of our own histories and prevailing meaning systems. The artist's rendering through portraiture cannot capture the whole essence of a person, nor can our exploration of the past represent any single truth. Just as the portrait both represents an interpretation and fails to capture the personality, struggles, or dreams of the individual being painted, so a rendering of the modern era must be considered interpretational and incomplete, while potentially, in its telling, transformative. As Mead (1932) suggests:

> There is an entire absence of finality in such presentations. It is of course the implication of our research method that the historian in any field of science will be able to reconstruct what has been, as an authenticated account of the past. Yet we look forward with vivid interest to the reconstruction, in the world that will be, of the world that has been, for we realize that the world that will be cannot differ from the world that is without rewriting the past to which we now look back. (p. 3) … I revert here to my original proposition that a reality that transcends the present must exhibit itself in the present. (p. 11)

Meaning and understanding are a function of language as our understandings and ideas are revealed through the language we use. Language, as the river of meaning through which we swim, is the medium through which meaning is supported. At the same time, the flow and quality of that water impacts the ecosystems that it supports, just as the richness of our language affects the depth of our ideas. Through metaphor, we create new meanings and reveal underlying ideas.

Lakoff and Johnson (1980) explain how metaphor goes beyond poetic imagination and is fundamental to our conceptual system. In order to create our curriculum futures, meaning must be made of our present curriculum situation and our curriculum past. Guiding metaphors of the curriculum can offer a perspective of how we have conceptualized the curriculum. Changing metaphors can also be important for directing or influencing our curriculum future and addressing the postmodern struggles of this new millennium.

This chapter will explore the standard modernist metaphors of schooling with their implications for curriculum theory and practice. As these metaphors are explored, the roots and characteristics of modernism, which serve as the foundation of the modern curriculum, will be traced. The philosophical and intellectual traditions of our times, including the major problems and approaches of philosophy, will be presented to further characterize modernity and present the *Erwartungshorizonten* of this book. The discussion in this chapter will lay the framework for interpreting the underlying logic of the modern agenda. The neoconservative education movement will be explored from within this context more fully in the next chapter along with the underlying logic of modernism, the logic of domination. Our "how" questions (How do we transform the curriculum? How can we change schooling?) require a recursion of "why" questions. Why are we as we are? Why does change seem so difficult? Why haven't previous reform efforts worked?

Guiding Metaphors

Kliebard (1975) discusses three metaphors that typically characterize approaches to the design of the curriculum. Production, growth, and journey metaphors pervade how we organize the curriculum and make sense of our goals for schooling. In particular, Kliebard (1998) suggests the production, growth, and journey metaphors characterize distinct values for education, defining goals, and having implications for practice. According to Kliebard, production, growth, and journey metaphors of education emphasize social efficiency, child study, and social reconstruction, respectively.

I have made it standard practice in many of my classes to ask students what their metaphors for schooling are. To avoid limitations of or biases introduced by language, I usually ask them to draw pictures to represent their metaphors. In every class, I get pictures of walls with bricks or buildings with foundations, gardens with flowers and sun and rain, and mountains with pathways to the top. When probed, students develop their metaphors with implications for teaching, learning, and knowledge. When further queried, they describe the roles of the teachers, students, administrators, parents, and society. They explore the purpose of schooling, as revealed through their drawings, and often add to their depictions to indicate the relationship between education and society. Sometimes their metaphors break down. While the path up the mountain seems like a nice representation of schooling, who determines the path? Where are the teachers? What is the mountain? Who decides?

As we discuss their metaphors, and as they are asked to describe what their metaphors imply about teaching, learning, knowledge, and the curriculum, what appeared as very distinct drawings often collapse into a common vision of schooling. Cummins and Sayers (1997) argue that perceived differences in the purposes of schooling and organization of the curriculum often are only appearances and do not represent fundamental differences or underlying philosophies. They suggest, for example, that the curriculum cycles and debates between content (traditional) focus and child-centered (progressive) approaches to the curriculum do not represent different approaches to curriculum inquiry but are merely "different sides of the same coin" reflecting Freire's "banking metaphor" (Freire, 1970).

> Education ... becomes an art of depositing, in which the students are the depositories and the teacher is the depositor. Instead of communicating, the teacher issues communiques and makes deposits which the students patiently receive, memorize, and repeat. This is the "banking" concept of education, in which the scope of action allowed to the students extends only as far as receiving, filing, and storing the deposits. (p. 58)

The banking metaphor represents the same technical, production interests as the factory metaphor described by Kliebard and underlies both child- centered/progressive approaches and content-driven, established approaches to the curriculum.

When asked about what we want our students to know, I too have found that students with building, growth, and journey models often say the same thing. They feel there is a body of knowledge that students should know both to benefit the student and to benefit society. While they prefer to view the teacher as "facilitator" rather than "sage on the stage," they nevertheless have clear ideas that the teacher should know more than the students and where the students should "end up." While they may recognize that their students are "different kinds of flowers or plants" or that the teacher needs to "provide different nutrients" to sustain their growth, it is clear the teacher has the power and is responsible for exercising that power in order to ensure appropriate outcomes.

Our ability to invent "new" metaphors for schooling may be limited by our own modernist assumptions, paradigms, and underlying logic. How do we learn to see things anew, have different visions, or even know the different questions to ask? How can I help my students see education from a perspective that is not driven by modernist assumptions and methods? Is the banking metaphor so ingrained in our common experiences of commerce and commercialism that we, at least in the West, cannot but help collapse all attempts to re-vision schooling into a banking model?

A result of approaches to the curriculum that have an underlying banking orientation is that schools become "irrelevant" to children and families, especially those disenfranchised from the dominant culture. These approaches also implicitly teach "dependence upon authority, linear thinking, social apathy, passive involvement, and hands-off learning" (Cummins & Sayers, 1997, p. 146).

Posner (1998) suggests that the "technical production perspective" of curriculum planning has been predominant over other approaches to the curriculum and often underlies approaches to curriculum reform. He notes that the predominant curriculum paradigm determines the kinds of questions, answers, and values that shape the educational landscape and the domain of curriculum inquiry. Within the technical production perspective:

> Schooling is assumed to be a process whose main purpose is to promote or produce learning. Students are termed *learners*; objectives are conceived in terms of desirable learning; evaluation of the school's success is targeted almost exclusively on achievement test scores; "educational" goals are distinguished from "noneducational" goals by determining if they can be attributed to learning, "curriculum" is defined...in terms of "intended learner outcomes." Thus, schooling is perceived as a *production system*, in which individual learning outcomes are the primary product. After all, if learning is not what schooling is for, then what could be its purpose? (pp. 81–82)

Sirotnik (1998) would concur that, although classrooms seem to have "become a bit more open" (p. 62), the "ideological domain" of the technical approach to the curriculum, guided by the factory metaphor, is still predominant. He explores the gap between the rhetoric of curriculum theory and curriculum practice, echoing what Goodlad and others have suggested, "additional domains must be described and interpreted if one wishes to engage in a thorough curriculum inquiry" (p. 60). Efforts to address the ideological domain of the curriculum have become more focused as the field of curriculum inquiry was born.

Traditional perspectives of the curriculum include curriculum as the content or subject matter to be taught, guides for what should be taught, lesson plans, learning goals, or actual classroom experiences (intended or unintended). The etymology of "curriculum" literally implicates the "racecourse" or plan of study. Each of these traditional perspectives defines the "racecourse" as the set of planned or experienced activities defining the purpose and product of school experiences.

Another derivation of "curriculum" implicating a different perspective is "currere" or "running of the race." While "curriculum" implies a set course run by all, "currere" suggests individual differences, goals, purposes, and values may guide our educative experiences. Considering "currere" rather than curriculum requires an exploration of individual histories and experiences.

Curriculum, as a discipline, has a variety of perspectives (see Schubert, 1986). Curriculum *history* "deals with processes of describing, analyzing, and interpreting past curriculum thought and practice" (Schubert, 1986, p. 41). Curriculum *development*, *design*, *implementation*, and *evaluation* determine what the "racecourse" should be, how to design it, how to get students to run it, and how to evaluate how they ran it. These perspectives of curriculum studies also have implications for teachers as well. What are their roles as the students run the race? How effective are they? How can they be more effective?

Curriculum inquiry and curriculum theory are relatively recent additions to the field of curriculum studies. Although these terms are often used interchangeably, curriculum inquiry assumes a critical or theoretical focus while curriculum theory includes philosophical perspectives.

Curriculum inquiry was initiated in the United States by the work of Elliot Eisner and set out as an agenda for curriculum studies by William Pinar in his address at the 1978 American Educational Research Association (Marshall et al., 2000). Curriculum inquiry has significantly changed the landscape of theoretical curriculum approaches. Critical and postmodern philosophies have grounded curriculum inquiry, laying bare the social, political, ethical, aesthetic, philosophical, and moral issues associated with our current education system and curriculum practices. Contemporary curriculum discourse as political, racial, gender, phenomenological, poststructural, deconstructed, postmodern, autobiographical, aesthetic, theological, and institutionalized texts has emerged as central to the field of curriculum inquiry (see Pinar et al., 1995). The modern agenda of efficiency, uniformity, and utility has been challenged by curriculum theorists by exposing how existing social, political, and educational structures perpetuate unequal power relations and by critiquing objectivity over subjectivity, single perspective over multiple perspectives, usefulness over fairness and justice, and homogeneity over diversity within these various discursive fields.

At the same time curriculum inquiry has been emerging, neoconservative political agendas are reestablishing curriculum frameworks, demanding uniformity of the curriculum, establishing national standards, and considering national testing to reestablish rigor and quality in public education. The lack of confidence in public education and a general feeling of impending educational crisis have loomed over the landscape of the American political agenda during these last two decades of the 20th century. As described by Marshall et al. (2000):

> From the ashes of the 1970's slowly arose the Phoenix of political conservativism to dominate the remainder of the century, effectively challenging progressive assumptions about the role of the state in social economic progress as well as individual and collective rights and responsibilities.... By 1983, the neoconservative revolution was evident to all: the publication of *A Nation at Risk* decrying poor test scores equated to a declining share of the global market.... [During this time] the proportion of mothers with children five years of age or younger working full- or part-time doubled to 47%, the proportion of heads of households between the ages of 18 and 24 who owned their home dropped by one-third, and the percentage of pretax income needed for mortgage payments on a median-priced home doubled. Meanwhile, the median American family income dropped 10% and unemployment rose to its highest level since 1941. (pp. 147–148)

Within this political climate, curriculum inquiry has taken a back seat to the more traditional approaches to curriculum studies. The perceived impracticality of the theoretical nature of curriculum inquiry, as well as the critical and questioning stance toward assumptions of rigor, purpose, and standards inherent in curriculum inquiry, have forced many curriculum theorists to back away from efforts to change schools and address problems from any other than a technical fix approach. These reconceptualist approaches to the curriculum have had little impact on curriculum practice in this political and social climate. The agenda of this new millennium within the curriculum field will be to engage in discourse with the neoconservative movement and to bring together curriculum theory, inquiry, and practice in ways that will make schools meaningful places for all children to learn. Challenges to meaning and evolving ideas about learning should ultimately transform our ideas about education and schooling.

In order to engage in meaningful and critical inquiry of traditional approaches to schooling and to address the neoconservative control over public education, the current production models of schooling will be explored from our historical and political roots in modernism. Modernist traditions and the need to find comfort in modernist paradigms drive the neoconservative movement and the desire for increased test scores, uniform curriculum, and successful international competition. In order to understand neoconservativism, or metaphors of schooling, and the impact of late 20th–century politics on education, we need a clearer understanding of modernism.

Roots and Characteristics of Modernity

There is little agreement about when the modern era began. As a significant period in the history of humans, however, this period can be distinguished by the ideals and meaning structures that contributed to the break from premodern times. Dramatic changes in social, political, economic, scientific, and technical cultures marked this transition from the premodern to modern world, suggesting a fundamental shift in worldviews pervading all areas of human endeavor.

The *zeitgeist* of modernism pervades all aspects of society. This shift in worldviews occurred throughout the 15th to 18th centuries. Several historical occurrences collectively can be identified that contributed to and reflect the changing meaning structures and ideals of Western society from premodern times. These changes, jointly, over time, are both the indications of and impetus for the advent of the modern era. As the middle-aged, over-weight writer may ask, "When did I get so fat?", the cumulative impact of a variety of historical events contributed to the transition from premodern to modern times. Some of the significant events impacting this transformation include:

- 1436 Gutenberg's adaptation of movable type
- 1520 Luther's rebellion against Church authority
- 1543 Copernicus' *On the Revolutions of the Celestial Spheres*
- 1630s Descartes' logic and epistemology
  Galileo's astronomy and mechanics
- 1648 End of the 30 years war
- 1687 Newton published *Principia*
- 1750 Development of steam engine and advent of the Industrial Revolution
- 1776 American Revolution
- 1789 French Revolution

From a political perspective, the modern mind-set, with its emphasis on the rugged individual and self-determination, can be traced to or evidenced by the American and French revolutions in the 18th century. These developments, of course, cannot be separated from the technological advances resulting from the invention of the steam engine in 1750 and the advent of the Industrial Revolution nor from the defiance of Luther and his proclamation of independence from the dominance of the Catholic Church.

From the perspective of scientific inquiry, the scientific method of Galileo, presented in his *Dialogues Concerning the Two Principal World Systems*, or Descartes' method of reasoning and emphasis on certainty as articulated in his *Discourse on Method*, both published in the 1630s, profoundly influenced philosophic and scientific thinking in the modern era. Later that century, Newton's *Principia* had a profound effect on the development and significance of mathematics. Together, the foundations of modern physics, philosophy, and mathematics were established in the 17th century by these thinkers. Their emphasis was on objectivity, theory, abstraction, rationality, and certainty, which are cornerstones of the modern mind-set.

Within the realm of religion, impacting the political and scientific worlds as well, the rejection of religious doctrine and subservience to the Catholic Church resulted in the formation of a variety of Christian religions and accompanying religious-oriented nation-states in the West. It might be argued that this religious revolution resulted from the invention of the printing press and the release of the first mass-produced version of the King James Bible. Even television's Arts and Entertainment national production of *Biography* named the invention of movable type to be the discovery of the millennium. Luther's rejection of church authority was another significant event in the rise of Protestant religions and beliefs that individuals could pursue and understand the "word of God" without the interpretation and authority of the Pope.

Collectively, the political, scientific, philosophical, religious, and technical advances of the 15th through 18th centuries have shaped the 20th-century perspective. From these origins, the basic tenets of modernism can be summarized as an emphasis on scientific reasoning and individual rationality, an assurance in universal truth, and a certainty in social progress (Fleener & Fry, 1998). Each of these cornerstones was formulated and solidified from all of the political, religious, scientific, and economic perspectives. These cornerstones of modernity are described below.

Individual rationality

The conviction that we can know anything of value or significance, for ourselves, without the authority of the church or interpretations of the Pope was first initiated during the reformation period. Because the printing press increased the availability of the King James' version of the bible, 15th- and 16th-century thinkers no longer needed priests to interpret the bible for them. Descartes and his method of philosophy formalized this independence of thought and belief that individuals could make reasonable judgments. Descartes' Cogito ("I think, therefore I am") emphasized individual human reason as the final arbiter of knowledge. His method of doubt, to only accept what is certain as foundational truth, became the paragon of reasoning in the modern era. As the basis of all knowledge, belief in the primacy of individual rationality empowered humans with the ability to reason and discover truths about the nature of reality. Mathematical relationships advanced by Isaac Newton imposed a uniformity, regularity, and predictability of planetary motion that gave further credence to the perspective that through rational deliberation, humans could discover the inner workings of nature.

The emphasis on the intellect in the modern era has ramifications for all aspects of social and intellectual life. The faith in the human intellect "celebrated the individual as opposed to established authority...the supremacy of reason, of the individual, and of individual freedom...Protestantism in religion, self-expression in the arts, (and) experimentation in science" (Elkind, 1995, p. 9).

Picturing the universe as a giant clock is the dominant metaphor for the modern perspective. Viewing the universe in this way implies that humans can understand the mechanics of reality as the jeweler understands the mechanics of a clock by taking the clock apart, studying the pieces, and putting it back together again. Mechanical time, in fact, became more than a metaphor for scientific inquiry during the Industrial Revolution. Time itself took on a new meaning within our social, economic, and political institutions (Capra, 1984). Mechanical time, unlike cosmic time, was to be controlled, measured, and further delineated into finer and finer increments. Like the clock-maker, in a mechanical universe, we have not only the means of understanding our reality by studying its pieces, but also the ability to define and fix it as well. As we pull our universe apart to understand its components, we are empowered to replace the broken parts and fix the clock to our specifications; thus, we are mandated to control our environment for individual, social, and political progress. With advances in medicine, even our own bodies were viewed as clocks. If a piece of our body-clock didn't work, we learned to fix it, remove it, or replace it to keep the clock ticking.

Implicit in the modern worldview, with its emphasis on individual rationality, is the belief that rational progress of society will contribute to more freedom for the individual. As it has evolved during the modern era, individual rationality became the basis for certain knowledge and was vital to freeing the individual from social, economic, and natural constraints. As argued by David Harvey:

> The scientific domination of nature promised freedom from scarcity, want and the arbitrariness of natural calamity. The development of rational forms of social organization and rational modes of thought promised liberation from the irrationalities of myth, religion, superstition, (and) release from the arbitrary use of power. (Hargreaves, 1994, pp. 25–26)

Power and repression could only be exercised against a body, not against a mind, from this perspective. Individual rationality can overcome the inequities and forces of tyranny and nature. Freedom is possible because we can think.

Social progress

The power of the intellect to both discover the inner workings of nature and free the individual from social and natural constraints, as described above, is intimately connected to the second cornerstone of the modernist perspective, namely the belief in social progress. Social progress is "the idea that society and the lot of individuals within that society are gradually improving" (Elkind, 1995, p. 9). The visions of rags-to-riches and examples of Horatio Alger and other 19th-century "self-made men" offered the promise in the United States that through hard work we as individuals could progress and better ourselves just as our Western society was advancing. The growth in scientific knowledge and advances in technological innovations during the Industrial Era are further examples of social progress as defining features of the modern era. Parents during much of this period, until the last few decades of the 20th century, have believed that each generation will have it a little easier, make a little more money, and make greater advances than previous generations.

Social progress is intimately connected with increased knowledge and individual rationality. Hargreaves (1994) summarizes this relationship:

> Modernity rests upon the Enlightenment beliefs that nature can be transformed and social progress achieved by the systematic development of scientific and technological understanding, and by its rational application to social and economic life. (p. 25)

Social progress, tied as it is to technological advances and individual rationality, is so ingrained in the Western modern mind-set, that as postmodern tensions occur, this aspect of modernism suggests a backlash to the past. Back-to-basics in education is a reaction to the challenge that society is not progressing, that our children are not smarter and better off than we were, and that society is deteriorating. We backtrack when we are lost, to rediscover our route. Those who would have us recapture the 1950s hope that, by returning to those times and values, we can continue our upward movement. Rather than challenging the underlying assumptions of modernism, these reactionaries are like the lost traveler who returns to the fork in the road, but tries the same route again. Those who cling to modernist ideas will only be able to return to the beaten path. Their certainty that there is a "right" path is a modernist inkling that there are given and determined truths to be discovered.

Universal truths

Scientific rationalism, technological and social advances, and the power of the intellect to discover the inner workings of the universe implicate a belief in universal truths. Scientific and social theories as articulations of universal truths predominate the modern perspective. By using objective, verifiable, replicable modes of inquiry, these truths, according to the modern perspective, are discoverable. Thus, through disinterested, value-free inquiry, universal truths can be uncovered.

The belief that, through rational, dispassionate, objective, replicable means, universal truths could be found, set a value on the kind of knowledge most worth having. No longer were artistic or even religious matters considered of supreme importance because there were no procedures by which certain knowledge could be obtained. During the modern era, in an attempt to counteract the dominance of religion and reliance on faith, our search for certain and universal knowledge significantly narrowed the field of inquiry. The over–emphasis on individual rationality, certainty, universal truth, and social progress has created a blindness or disequilibrium that worships scientific rationality over all other forms of reason.

Toulmin (1990) argues that, rather than representing a time of great advance, the 17th century and the advent of Cartesian Rationalism was a time of narrowing. He suggests that the Renaissance humanism of the 16th century more adequately suggests a break with medieval thinking and that the rise of science and philosophy in the 17th century in effect fragmented the perspective of Renaissance humanism into two cultures. He describes:

> The contrast, between the practical modesty and the intellectual freedom of Renaissance humanism, and the theoretical ambitions and intellectual constraints of 17th century rationalism, plays a central part in our revised narrative of the origins of Modernity. By taking the origin of Modernity back to the 1500's, we are freed from the emphasis on Galileo's and Descartes' unique

> rationality.... If the Two Cultures are still estranged, then, this is no local peculiarity of 20th century Britain: it is a reminder that Modernity had two distinct starting points, a humanistic one grounded in classical literature, and a scientific one rooted in 17th-century natural philosophy. (p. 42–43)

The fragmentation of modernism represented by these two movements is keenly felt as we enter the 21st century. Dualisms (mind/body, theoretical/practical, male/female, objective/subjective, science/humanities), language, meaning, and change are major areas of concern as we struggle with the consequences of our own intellectual traditions. The lack of balance in society today is perpetuated in the curriculum where science and mathematics are more highly valued than the arts, literature, or language.

Voices of dissent have been heard, though virtually ignored, throughout the modern era. The romantic poets and scholars spoke in direct opposition to the Rationalist movement. Blake, Goethe, Wordsworth, and Emerson echoed the concerns of Rousseau, Kant, Hegel, Marx, and Engles. Focused on organic, holistic, and relational metaphors, objections to the mechanistic perspective of modernism are finally being heard. While the technical and scientific advances have been staggering, there is a sense in the early 21st century that we have lost our balance and are careening into a crevice from which we cannot escape. In our quest for security, we have lost our freedom. As we have followed our technological gods (Postman, 1995), we have become prisoners in a prison without bars (Foucault, 1977).

A Loss of Balance and Voices of Dissent

From the perspective of Western philosophy, the overriding problems of philosophy have been accounting for the relationships among abstract ideas like mathematics or colors, our belief in an independently existing external world, and accounting for our experiences, including experiences of motion, change, creativity, and free will. Since Descartes (1596–1650), the questions of objective reality, the existence of other minds, and the abstraction of mathematics have been couched in mind-body dualism. His method of doubt, to reject all statements of truth except those that cannot be doubted, led Descartes to the foundations of knowledge resting in our awareness of our own thoughts. To build his epistemology on this most basic experience, "I think, therefore I am" and reestablish knowledge of the external world is problematic for Descartes. By basing all knowledge on our awareness of our own thoughts creates a separation of minds or thoughts and bodies external to our thoughts. As described by Whitehead (1925):

> Descartes discriminates two species of entities, matter and soul. The essence of matter is spatial extension; the essence of soul is its cogitation....Those principles lead straight to the theory of materialistic, mechanistic nature, surveyed by cogitating minds. After the close of the seventeenth century, science took charge of the materialistic nature, and philosophy took charge of the cogitating minds. Some schools of philosophy admitted an ultimate dualism; and the various idealistic schools claimed that nature was merely the chief example of the cogitations of minds. But all schools admitted the Cartesian analysis of the ultimate elements of nature. (p. 145)

Descartes' emphasis on matter and motion as the essential attributes of nature (Kline, 1982) and Newton's metaphysics of physical entities, located in space and time have predominated scientific thinking since their inception. The Cartesian method of rational thinking and the scientific rationality of Newtonian physics have been the paragons of epistemology for the last three hundred years. This break with Aristotelian thought in the 17th century, with an emphasis on epistemology over metaphysics, was grounded in the quantification of nature (Crosby, 1997) or the mathematization of science (Kline, 1982). As explained by Kline (1982):

> Descartes...discarded all these qualities and insisted that all physical phenomena could be explained in terms of matter and motion. The essential attribute of matter was extension and this was measurable and so reducible to mathematics. Further, there could be no extension without matter. Hence a vacuum was impossible. Space was filled with matter. Moreover, matter acted on matter only by direct contact. (p. 51)

The quantification of nature and the emphasis on causal interactions of inert matter set the agenda for modern science and provided a mechanistic view of the inner workings of our universe. This mechanistic perspective provided the fertile soil for an underlying logic of domination, discussed in the next chapter. Mathematics played a major role in the practice of science as well as the justification for scientific theories. If relations in nature could be mathematized, and because mathematics is perceived to be Absolute Truth and objective, then our theories must be correct. As described by Kline (1982):

> In retrospect one can see that the doctrine of God's mathematical design of nature was being undermined by the very work of the mathematicians themselves. The intellectuals became more and more convinced that human reason

> was a most powerful faculty and the best evidence for it was the successes of the mathematicians. (p. 73)

Many of our commonsense notions of average, aggregate, and even space and time come from developments in Renaissance and 17th-century mathematics. The idea that the average family has 2.5 children would make no sense to someone living during the Middle Ages. Similarly, to represent the motion of an object as a time-distance continuum on a Cartesian graph has tremendous implications for the relationship between time and space, and motion and change. Mathematics became in the modern era the "language of the Gods" or the language in which nature's secrets were written. Mathematics thus is the tool by which, through scientific rationality, all knowledge could be discovered and verified.

Thus, since Descartes, the legacy of epistemology has been how to account for knowing minds that seem separate and distinct from the inert material world. Modern philosophy has had difficulty addressing the consequences of the modern scientific view of reality, specifically with regard to identity, dualisms (mind/body, subject/object), change, creativity, freedom, time, relativity, quality, process, and power (Lucas, 1983; Reck, 1975). For example, in line with Newtonian mechanics, Spinoza (1632–1677) viewed the natural unfolding of nature as mechanical, deterministic, and therefore predictable. Although addressing many of Descartes' difficulties, in so doing, Spinoza introduced other problems, including the problems of free will, creativity, imagination, spirit, and experience. Both Spinoza and Leibniz after him, to avoid reducing human nature to that of an automaton, infused spirit into the fundamentally mechanical universe. The need to do so points to a flaw of dualistic thinking. Without such a move, the implications of the modern worldview are that either our experiences of free will and creativity are unaccountable or questions of the existence of an objective, knowable, external reality, fundamental to our modern science, become problematic.

While philosophy in the 18th and 19th centuries continued to address these concerns, society left philosophy to the philosophers, separated science from philosophy, and approached the technical functioning of scientific and social inquiry from unquestioned theoretical roots in dualism. In opposition to the rationalistic perspective were the Romantics of the 18th and 19th centuries. Romanticism offers in its opposition to rationalism a perspective of modernism as well as potential alternatives to it.

Romanticism

Voices of dissent, still cited today, Vico, Blake, Goethe among others, arose not in unison but in a cacophony of sounds, uniform in their rejection of rationalism, the separation of humans from nature, the mechanistic perspectives of the universe, and an emphasis on universal truths. "We murder to dissect," complained Wordsworth; "Are not the mountains, waves and skies a part of me?" asked Byron. The knower was left out of Descartes' equation, yet, according to the Romantics, a sense of the known entails a sense of ourselves. If we remove ourselves from that which is known, we lose sense of that which is meaningful. We cannot separate the meaning of things from discovering our own meaning and, ultimately, in the name of emancipation from an over-bearing church, our emphasis on technical science, rational thought, and mechanistic control have cost us our freedom. As described by Wilshire (1968):

> It dawned on the romantics that scientific knowledge with its power over nature had been used by man to change the world before he had determined what his real interests were; as if he had wielded a fabulous weapon but, unaware of the position of his own feet, had begun to hack them off. (p. 11)

Empirical and rational movements in philosophy could not escape the consequences of their own method. Meaning and purpose are not a part of the Newtonian world or of Cartesian doubt. From our perspective of *Erwartungshorizonten,* understanding how Western society became so fragmented, and listening to the voices of the Romantics will provide a perspective from which our visions of the future may emerge.

A common theme of many of the Romantic thinkers of the late 18th and 19th centuries was that parts or pieces of a whole are meaningful precisely and only because they are parts of a whole. To understand the processes of anything complex, especially living creatures, we must understand their relationships, purposes, goals, and understandings. Indeed, we must understand something about ourselves as we search for understandings about others. Life is an art and living is the practice of art. Our natural environment is crucial to our well-being and we must act in accordance with rather than in opposition to nature. Living in harmony with nature and feeling our connections with and dependence upon the earth, are perspectives of Romanticism that are ignored from the scientific rationalism of Descartes and Newton.

> I live not in myself, but I become
> Portion of that around me; and to me
> High mountains are a feeling. (Byron)

Meaning, purpose, and interconnection, rather than knowledge and control, motivate the Romantics to seek beyond themselves to gain an understanding of self and to feel and live life. At the same time, self-knowledge is a prerequisite to knowledge about our world.

The Romantics rejected the logic of modernism and embraced feeling, contradiction, paradox, and perspective. The dialectic, rather than rational thought, became the tool of reasoning for the Romantics, not toward the end of certainty, but for transformative and process-oriented purposes. The Romantics, however, had little influence outside their social sphere. Their voices of dissent have for almost two centuries been little more than a murmur.

The voices of dissent emphasizing interconnectivity, holism, relationship, meaning, and purpose are now being heard during these early years of the 21st century, however. The postmodern tensions resulting from the meaninglessness inherent in modernist mechanism and fragmentation are finally being experienced and explored by more than the poets, philosophers, and intellectuals. After World War II and especially apparent during the Viet Nam "conflict," the cornerstones of modernism have been crumbling away to nuclear fission, computers, and poverty rates on a worldwide scale never before seen.

Another voice of dissent, perhaps also not immediately understood or effective but relevant to our *Erwartungshorizenten* is that of the idealists. While many of the romantics were influenced by the idealism of Kant, Hegel, and Marx, it was not until theories of evolution and relational sciences in the 20th century that romanticism and idealism could together confront and provide an alternative perspective from which a postmodern worldview could emerge.

Idealism

Idealism was a philosophical movement of the 18th and 19th centuries created to address the problems of Cartesian doubt. Philosophers of science and epistemologists interested in how we know truths about nature, however, have virtually ignored or discounted idealsim. Some branches of Idealism reject that the external world exists. Others only go so far as to claim that, although the external world may exist, we cannot know anything beyond our ideas or experiences of it. Von Glassersfeld (1995) and some branches of constructivism reflect the perspective offered by Bishop Berkeley (1685–1753) who argued against the mechanistic approach of Newtonian physics and against Locke's empiricism by showing that the logic of their arguments naturally led to contradictions or unacceptable premises. Radical constructivism emphasizes that all we can know is our own sensory experiences and knowledge of any external world is constructed knowledge validated by our own ideas about consistency and constancy of experience. Radical constructivism is considered postepistemological precisely because it rejects epistemology or the creation of knowledge as a goal of its methods. If all we can know is our own sensory experiences, it makes sense not to pursue knowledge but our ideas about knowledge. Radical constructivism adopts this postepistemo- logical stance to derive meaning and purpose rather than certainty and foundations.

These and many other postmodern attempts to address the problems of modernism reinvent the arguments of idealism. Berkeley, in sections 23 and 24 of his *Principles of Human Knowledge,* written in the early 19th century, showed the mind's reflection on ideas (as sense data) about reality cannot be equated with or used to substantiate a separate existence to that reality. We can have no knowledge of anything other than our own ideas.

> All phaenomena (sic) are, to speak truly, appearances in the soul or mind; and it hath never been explained, nor can it be explained, how external bodies, figures, and motions, should produce an appearance in the mind. These principles, therefore, do not solve, if by solving is meant assigning the real, either efficient or final causes of appearances, but only reduce them to general rules. (As quoted by Smith & Grene, 1963, p. 2)

Immanuel Kant (1724–1804) also attempted to address the problems inherent in Cartesian doubt and dualism. He is considered by many to be one of the greatest and most influential thinkers of modern philosophy. Some view his approach as idealist while others feel his focus on reasoning places him within the rationalist perspective. As described by Moore (1990):

> Kant wanted to accept, with the rationalists, that we had substantial a priori knowledge. Yet, in line with the empiricists, he did not see how we could know anything substantial about what was out there, independent of us, without letting it impinge on us through experience. He resolved the apparent conflict here by arguing that the a priori knowledge in question was not after all knowledge about what was out there, independent of us. (p. 84)

As the "grand moderator" of the philosophical debates of the times, Kant's *Critique of Pure Reason* attempted to ground Newtonian mechanics while going beyond or transcending the limitations of our own sensory experiences. As with Berkeley, Kant challenged the assumed givens of science. His critical theory embraced the problems of Hume and suggested the problems of empiricism could be solved if we could show how knowledge could be derived from experience. Accepting that there are limits to our knowledge and that we cannot get beyond knowing our own sense perceptions, he developed categories of reasoning. Rather than investigating reality, his agenda was to study these categories of reasoning.[1]

Mathematics, for Kant, was an example of the powers and structure of our mind and was *synthetic* insofar as mathematics could offer us a special kind of knowledge of the physical world as experienced through our sensations. Because mathematics is *synthetic*, the knowledge offered is not about reality as such, but about our ideas of reality. Thus, Kant rejected that mathematics and the laws of nature quantified through mathematics were about external reality. This perspective is in direct contrast to that taken by the Newtonian agenda.

> His doctrine that what mathematics asserts is not inherent in the physical world but comes from man's mind should have given pause to all mathematicians.... Unlike Kant, mathematicians and physicists still believed in an external world subject to laws independent of human minds. The world was rationally designed and man merely uncovered that design and used it to predict what would happen in the external world. (Kline, 1982, p. 77)

Hegel expanded the tradition of German idealism, started by Leibniz and more fully articulated by Kant, in the form of his dialectical phenomenology. Consistent with the idealist tradition, Hegel (1770–1831) rejected that the world is made up of "things." From Kant he inherited and embraced the inherent conflict of reason with itself as it engaged in the dialectic with reason. Reason attempts to extend categorical understanding beyond understanding itself and apply it to things. Rather than accept the limitations of Kant's idealism and ultimate skepticism with regard to the existence of an external world, Hegel focused on the dialectical and evolutionary process of ideas themselves. The constitution of reality for Hegel is exactly this complex process, the evolution and interrelatedness of our own dialectic with nature. Self-consciousness likewise becomes for Hegel the evolution of our ideas about self through the dialectical process between self and other. Thus ideas about self require the conscious self to be aware of itself as "other" in other selves' deliberations.

With Hegel we see a shift from a discussion of ideas to a discussion of language as key to the dialectic process. The goal for him was therefore not to extend the reach of reason to overcome our own subjective experience, but rather to understand our reality as we experience it through our ideas. The Absolute reality, for Hegel, was spirit or mind in the process of self-actualization. The process of actualization was the dialectic through the resolution of conflict. The Absolute, according the Hegel, has an ultimate goal. As it evolves through its own historical development, the Absolute manifests itself in nature and in mind through human history. So as our reason evolves, the Absolute evolves.

Hegel's dialectic phenomenology influenced social theory (Marx and Engels), existentialism (Kierkegaard), critical theory (Lacan, Habermas, Gadamer), process metaphysics (Whitehead), and process theology (Hartshorne). Many of the analytic philosophers at the beginning of the 20th century (Bertrand Russell and G.E. Moore) felt the influence of Hegel to be so great that his theories specifically needed to be countered and were addressed by them. The significance of the Idealists, from an historical perspective, as well as their importance for later arguments in this book, are marked by their break with the traditions and focus of science in 17th- and 18th-century philosophy and reflects the emphasis of the Romantics on the dialectic, feelings, interconnectivity, and transformation. The Idealist and Romantic traditions are included as part of the presentation of our intellectual heritage. They also reveal the tensions that have led us to our own needs to resolve the problems inherent in the modernist intellectual tradition. Creativity, conflict, language, relationship, and process become important considerations for a postmodern curriculum and are explored in later chapters.

The next chapter will further explore the problems of the *zeitgeist* of modernism by considering the underlying logic of modernity. In so doing, the political and social ramifications of the modern mind-set will be explored. The gain in momentum of the neoconservative movement in the last two decades of the 20th century will be explored using the context of our foundations in modernism and the underlying logic of domination and scientific rationality. Later discussions of postmodernism will revisit the philosophical debates and attempted solutions to the problems of Cartesian Doubt and the consequent dualisms that have bifurcated our society. The whys of our questioning will assume a different nature as we proceed in our exploration of curriculum futures from the perspective of *Erwartungshorizonten.* Rather than the definitive "because" of the exasperated parent, we will consider the chain of whys as a process, holistically painting a picture of meaning, feeling, and experience.

Notes

1 Kant distinguished between things-themselves (Ding an Sich) and things as we experience them (Noumena). Recall that, with Descartes, if all we can know is our own mind, a chasm is created between our minds and the material world. Kant tried to address this problem. His goal was to determine how knowledge of ideas beyond our experiences, eternal knowledge, was possible. To do this, he distinguished between *analytic a priori* knowledge and *synthetic a priori* knowledge. *Analytic a priori* knowledge is the result of logical analysis. Some mathematical knowledge, the truth of a theorem in geometry, for example, is *analytic a priori. Synthetic a priori* knowledge is the knowledge inherent to the structure of our minds. According to Kant, our ideas about mathematics are not derived from our generalizations (synthetic judgments) of experiences with space (geometry) and time (arithmetic) but are *a priori synthetic* judgments or knowledge inherent to the structure of our minds. As such, our ideas about mathematics are given before experience (a priori), but originating from the inherent structures of our minds (synthetic) rather than from their own internal logic (analytic). Categories of *synthetic a priori* knowledge include notions of unity, plurality and totality, reality, negation and limitation, inherence or subsistence, causality or dependence and community, possibility or impossibility, existence or nonexistence and necessity or contingency.

Synthetic a priori judgments, although ideas, are related to experience. As explained by Smith & Grene (1963):

> *A priori synthetic* judgments are possible, according to Kant, only in so far as actual experience necessarily presupposes them. They are therefore valid only in relation to experience. (p. 259)

Unlike *analytic* propositions, which can be analyzed deductively, *synthetic* propositions allow for new knowledge. Thus, although knowledge may begin with experience, the structures of the mind allow for the generation of new knowledge.

CHAPTER TWO
The Logic of Modernism

I have taught only one student in my entire career as a teacher with whom I never found a common ground. I know we all have had students like her, but I always felt I just could have done more.

She used to come into my mathematics class, loud and aggressive from the moment she entered the door. She'd slam her books on her desk and ask in an accusatory voice, "Are we going to learn anything today?" During class she'd roll her eyes, sigh so loudly that the tobacco factory workers downtown could hear her, and flay her arms about in exaggerated displays of frustration.

As a first-year teacher, I tried all of the textbook approaches to managing her behavior. I tried reasoning with her, isolating her, giving her alternative assignments, making note of her behavior in front of the other students, discussing her behavior in private, giving her "teacher looks," sending her to the counselor, sending her to the principal, reporting her to her JROTC instructor…everything! Nothing worked. Finally, at the end of the first semester, there was a mutual agreement among the principal, student, and myself to have her drop the class.

Looking back, I realize I approached her from a position of dominance and authority. We immediately engaged in a power struggle that neither of us could win. Our relationship reminded me of the cold war with occasional outbreaks of violence. More usual, however, were the occasions of dirty looks, and disrespectful or disparaging comments. Seething below the surface were distrust and dislike apparent in all of our interactions. Somehow, we never connected as human beings, as individuals who had needs, desires, and our own private motivations.

This experience was important for me as a later university methods instructor. Teaching methods classes has always presented a problem for me because the effectiveness of methods of classroom management has nothing to do with the methods themselves and everything to do with the relationships and feelings of trust and care shared by all.

The logic of domination and underlying metaphors of schooling pervade all aspects of the classroom environment including how we interact with our students and approach our instruction. Even though I'd never want to go back to being a first-year teacher, I have thought over the years how I would have changed my interactions with that student. Control, organization, order, management are all words that reflect a modernist perspective. Replacing these ideas and rejecting the underlying logic of domination with postmodern logics of relationship, systems, and meaning, I would approach this student very differently now. This chapter will explicate the logic of domination and explore the implications of this underlying logic for classrooms and schools. Later chapters will develop the postmodern logics with their implications for schooling.

Modernism has been characterized in many ways. The previous chapter considered the multiple influences and origins of modernity (social, political, economic, scientific, and religious), delineated the tenets of modernism, and traced the major philosophical problems and approaches of the modern mind-set. The voices of dissent, rising above the clamor of the technical approaches, were presented as vantage points from which our own postmodern perspectives may emerge.

The focus during modern times on epistemological method and ontological certainty define a social meaning system that is quintessential Modernistic. Ontological and epistemological perspectives are foundational to all social meaning systems and influence how we perceive being, space, time, relationship, creativity, knowledge, power, social structure, spirituality, change, and causality.

The modern perspective has implicit in its system of thinking an underlying perspective of the quantifiable and intimately connected natures of space and time resulting from the predominance of the scientific paradigm and ideas about the rationality of humans. Developments in mathematics will be explored as essential to the scientific paradigm and underlying logic of human rationality characterizing the modern mind-set. Implications of the mathematization of reality include an underlying logic of domination and mechanical perspective of the universe that have direct impact on what counts as knowledge. These underlying logic and mechanistic perspectives similarly have an impact on the ethical and moral milieu of the era, which will also be discussed, especially in relation to schooling and the curriculum hierarchy. Finally, notions of space, time, and object inherent in the modern mind-set will be examined to compare with 20th-century scientific revolutions and our evolving ideas during our postmodern transitions.

Our exploration in this chapter will begin with an examination of the rise in the late 20th century of the neoconservative movement. The neoconservative movement and its influence on education are reflections and indications of the problems of modernism. Current curricular practices that have at their roots the neoconservative agenda also reflect the underlying logic of modernism. The perception of an impending crisis in education is both a result of the underlying tenets of modernism and a ploy to solicit action that perpetuates the neoconservative agenda.

Neoconservativism and Curricular Crisis

Fletcher (2000) delineates the neoconservative perspective in education as follows:

> What distinguishes neo-conservative views on education is their promotion of falsely objectified and depoliticized disciplinary knowledge at the expense of diverse student experience, and a monolithic set of exclusionary cultural values at the expense of critical reflection. (p. 11)

The roots of modernism can be traced within these two key features of the neoconservative agenda, namely a focus on disciplinary knowledge and a critique of experience. These roots are reflected in the platform of the neoconservative education agenda. According to Fletcher, the neoconservative curriculum agenda specifically challenges a multiperspective approach to the curriculum, challenges the value of individual human experience, denies that the American dream is inaccessible to some, and concludes that what is great about the American culture is that it is monolithic rather than diverse. His discussion points and the parameters of the neoconservative movement as it is expressed in public education are summarized below:

- Attention to experience and interest in schools is an obstacle to learning valuable content.
- Attention to experience and interest in schools is an obstacle to sound moral development.
- Rather than a curriculum based on individual interest and experience, there should be a common curriculum for all.

- The curriculum in American schools should be based on American culture.
- Multicultural education is based on an attempt to politicize the curriculum.
- Multicultural education is an obstacle to integrating minority students and breeds divisiveness and separatism. (Paraphrased from discussion points in Fletcher, 2000, pp. 12–22)

While the modern American culture was built upon the tenets of modernism with foundations in individual rationality, social progress, and universal truths, individual experience per se takes a back seat to a special kind of rationality-scientific rationality. Scientific rationality is responsible for the technologic, economic, and social gains of the 20th century and reveals, from the modernist perspective, universal truth. Individual accomplishment is measured in terms of success within this rational structure. Objectivity is valued over subjectivity as a reflection of the process of scientific rationality. Underlying scientific rationality and the modernist perspective is a logic of domination that can be traced back to the origins of our mechanical ideas about how the universe operates.

Origins of Scientific Rationality and the Mathematization of Reality

Our perceptions of time and space changed with the invention of the mechanical clock in the Middle Ages. In particular, the inventions of mechanical time and canons, also invented in the Middle Ages, contributed to ideas about the quantifiable and intimately connected natures of space and time. For example, to get the maximum distance from a canon, one can mathematically calculate and quantifiably relate the angle of elevation, initial velocity, and distance traveled. As intuitive as this seems to us today, formalizing these relationships in a mathematical formula requires a perspective of quantifiable relationships among space, objects in motion, and time.

Time itself, with the invention of the mechanical clock, took on new meaning and facilitated the development of the perspective of a quantifiable universe. Mechanical time introduced new notions of control and independence. For example, with the invention of the mechanical clock, commerce and the behavior of workers could be controlled. As described by Crosby (1997):

> In 1314 the city of Caen installed a clock on a bridge and inscribed it thus: "I give the hours voice/To make the common folk rejoice." ... A fifteenth century petition for a city clock for Lyons proclaimed, "If such a clock were to be

made, more merchants would come to the fairs, the citizens would be very consoled, cheerful and happy and would live a more orderly life. (pp. 76–77)

Philip VI recognized the utility of clocks for commerce in 1335. The mayor and alderman of Amiens requested a clock to regulate their workers. Philip VI agreed and ordained the use of the town's clock to "control by a bell the time when the workers of the city should go to work in the morning, when they should eat and return to work after eating, and when they should quit work" (p. 86).

Clocks were most likely invented, according to Crosby (1997) by monks who used them to determine when to rise for their predawn morning prayers as well as to determine prayer times throughout the day. Without the mechanical clock, the monks had to rely on the sun and stars to tell the time. Inclement weather or darkness of night made accurate determination of time difficult. The invention of the mechanical clock therefore had the effect of making the monks independent of nature. As Crosby (1997) explained, "time began, for the first time in history, to be "isolated" as a pure form, exterior to life. Time, though invisible and without substance, was fettered" (p. 82). The invention of the mechanical clock in effect gave us control over time, or at least our use of it, and isolated time from nature.

Mechanical clocks also gave us ways of quantifying our universe. We could calculate the period of a pendulum swing or the time it took an object to fall from a tree. This mathematization of changes in physical phenomena through space and time was significant and captures an essential ingredient of the rise and influence of science during the modern era. By mathematizing these relationships we can make predictions about future events and inferences about causal relationships and change. The uncertainty of causality and forces of change were demystified as we developed the analytic and mathematical tools for capturing these relationships. Medieval mysticism and qualitative interpretations of time were replaced in the 16th and 17th centuries with quantitative, numerical notions of time.

The invention of the mechanical clock and the acceptance of the perspective of mechanical time provided a necessary foundation for the modern scientific paradigm. The relationship between objects moving in space and time and the mathematics that captured that relationship, fundamentally changed how Western societies approached nature and conceived of the potential of science to understand the inner workings of nature. As Fraser (1975) summarized:

> Between the demise of the Middle Ages and the birth of Kantian critical philosophy, one may observe two simultaneous processes in the intellectual history of time: the secularization of the idea of historical continuity and change, and the birth of its formal corollary, the mathematization of natural phenomena progressive in time. (p. 30)

The mathematization of nature and the development of the empirical sciences with an emphasis on rational procedures and objective measures through experimentation are the predominant influences of the modern mindset.

Galileo and Newton formalized the mathematization of reality in the late 16th and early 17th centuries with their developments in celestial mechanics. As described by Husserl (1997/1970):

> Its rationalism soon overtakes natural science and creates for it the completely new idea of *mathematical natural science*—Galilean science.... As soon as the latter begins to move toward successful realization, the idea of philosophy in general (as the science of the universe, of all that is) is transformed.... [T]hrough Galileo's *mathematization of nature, nature itself* is idealized under the guidance of the new mathematics; nature itself becomes ... a mathematical manifold. (pp. 22–23)

Every algebra teacher tells the story of how Descartes invented the coordinate system while lying on his bed watching a fly buzz around on the ceiling. He concluded that the exact location of the fly on his ceiling could be determined, so the story goes, by counting the number of squares down and the number of squares across the edges of the ceiling. The unique mapping of location in space of an object was crucial in the developing mathematization of nature and the development of calculus and physics. It also was instrumental in our developing and accepting conceptions of dimensionality, which have gone unchallenged until very recently.

Fraser (1975) further elaborates to show how the methods of modern science came to rely on mathematics and abstraction, removing the perspective of the observer. By removing the observer, as with the objectification of the curriculum, individual experience can be ignored. Just like Descartes' method of doubt, the goal was to find certain truths that did not rely on individual bias or positionality.

> The total separation of time both from the equations of motion and from the contingencies, including the observer himself, is the essence of Newton's dictum that "absolute, true and mathematical time, of itself, and from its nature, flows equably without relation to anything else."... The revolutionary contribution of the Newtonian synthesis to our understanding of temporality resides,

therefore, not primarily in the idea of absolute time but rather in the separation of being and becoming in the nature of physical time, permitted by that idea. (p. 33)

With the mathematization of science, the removal of context, and the focus on being, scientific methodology fundamentally changed. Scientific methodology became both empirical/testable and abstract/mathematical. Formal relationships, represented mathematically, could be explored. Mathematics became both the mode of abstraction and the means of predictive calculation and confirmation. Because of the mathematization of reality, the world, so perceived, and all of nature became quantifiable, predictable, and controllable.

> By means of pure mathematics and the practical art of measuring, one can produce ... a completely new kind of inductive prediction; namely, one can "calculate." (Husserl, 1997/1954, p. 33)

Science, before the modern era, was primarily influenced by Aristotle's classification approach. The emphasis on measurement and the mathematization of nature marked a separation from the emphasis on Aristotelian classification. The adoption of the mathematization of nature during the 17th century propelled modern science and the development of modern technologies. While not particularly surprising to us, the notions of measuring naturally occurring relationships and quantifying time are unique to the modern mind-set. As distinguished by Harding (1986):

> Not all cultures have been as preoccupied with measuring—dividing a line—as has ours for the last few centuries. A whole field of mathematical inquiry was made possible by the substitution of a different kind of social image for thinking about what numbers are. (p. 50)

Griffin (1980) outlined the various late medieval and Renaissance thinkers who developed the techniques and inferred this change in perspective to the predominance of measurement. Starting in 1382 with Brawardine's *Treatise on the Proportions of Velocities in Moving Bodies,* she traces the influence of measurement and the change in perspective from Bacon's *Natural and Experimental History for the Foundation of Philosophy* published in 1622 and Galileo's publication of *Two New Sciences* in 1638, to Newton's *Principia* published in 1687. She summarized these developments as follows:

> And it is said that nature can be understood only by reduction, that only by reducing her to numbers does she become clear.... It is decided that that which

> cannot be measured and reduced to number is not real. It is questioned whether or not motion is real. It is discovered that motion can be measured by measuring space through which movement moves and the time in which the moving takes place. It is decided motion is real.... It is said that God is a mathematician. (pp. 11–14)

With emerging techniques in calculus and algebraic symbolism, the notion of function developed. Being able to write formulas that seemed to capture the causal relationships between the motion of entities through time using functional notation seemed to be evidence that there were hidden truths or laws in nature that could be discovered through the scientific process and modern mathematics. These truths revealed, according to this thinking, an underlying causality and mechanism that was reflected by mathematical functionality. According to Whitehead (1925/1967):

> [T]his dominance of the idea of functionality in the abstract sphere of mathematics found itself reflected in the order of nature under the guise of mathematically expressed laws of nature. Apart from this progress of mathematics, the seventeenth century developments of science would have been impossible. Mathematics supplied the background of imagination through which the men of science approached the observation of nature. (p. 30)

The formalization of mathematics through formulaic and functional development and advances in algebraic representation further contributed to the mathematization of nature. Galileo's *hypothesis* that through mathematics we could explore naturally occurring relationships was not intended to suggest, however, that there was a hidden order captured by that very mathematics. Mathematics, originally a way of modeling our ideas about nature, soon became mistaken for an exact representation of the inner workings of nature. Throughout the modern era, humans, through scientific reasoning and mathematical prediction, have used mathematics to represent and confirm the belief that there are underlying laws of nature knowable. As described by Husserl (1997/1954):

> The indirect mathematization of the world, which proceeds as a methodical objectification of the intuitively given world, gives rise to general numerical formulae which, once they are formed, can serve by way of application to accomplish the factual objectification of the particular cases to be subsumed under them. The formulae obviously express general causal interrelations, "laws of nature," laws of real dependencies in the form of the "functional" dependencies of numbers. (p. 41)

The mathematization of reality resulting from the development of the mechanical clock and advances in mathematics has led to a separation of humans from nature, a presumption of control over and knowledge of the inner workings of the universe, the metaphor of a clockwork universe becoming mistaken for a representation of nature, expectations of predictability, confirmability, and certainty, and a functional relationship between entities in space and time. Underlying the mathematization of reality is the logic of domination. This underlying logic of modernism is foundational to and a consequence of modern scientific methods and subsequent deification of science as the paradigm of inquiry. The next section will describe the relationship between the mathematization of nature, the scientific paradigm, and the logic of domination that pervades the modern mindset.

Logic of Domination

The curriculum hierarchy both reflects the importance of the mathematization of reality as a fundamental paradigm of modernity and emphasizes the consequences of the conceptual framework upon which the scientific paradigm is based. Students who do well in mathematics and science are often considered brighter than students whose talents lie in the humanities or arts. Advanced tracks in mathematics, usually starting in middle school or junior high, often identify or label the "gifted" students in the school based on their mathematical abilities and achievements. In times of perceived national crisis, the route to societal and technological superiority is identified as coming from enhanced mathematics and science instruction. The neoconservative cries for educational superiority emphasize being "first" in mathematics and science. This perspective is especially apparent in the educational reform efforts resulting in New Math after Sputnik in the late 1950s and again with President Bush's Goals 2000, which emphasized a "number one" ranking in mathematics and science for students in the United States by the year 2000. President George W. Bush's cries for mandatory national testing in mathematics and reading reflect the perceived importance technical competency in mathematics has for national security, competition, and superiority.

In the late 1960s, mathematics education researchers noted the differential participation rates and achievement of girls in mathematics. In her study of college freshmen at Berkeley, Sheila Tobias identified and delineated the career limitations and potential for high-paying jobs resulting from young women's lack of mathematics preparation upon entering college. This research raised questions about why girls seemed to have lower achievement and less interest in mathematics- and science-related subjects and careers. As with other violent crimes against women, much of the research seemed to put the blame on the characteristics of the victims. Some research suggested a "math gene," or the natural capacity of men to reason mathematically, is lacking in women. Some argued hormonal changes occurring in adolescence when girl's performance and participation in mathematics and science wanes was a factor and that estrogen is related to artistic rather than "logical" thinking. Much of the left-brain/right-brain research similarly characterized women as right-brained, divergent thinkers rather than left-brained logical thinkers. Others have suggested peer pressure and lack of support from their families and teachers account for the differential between boys and girls in mathematics and science achievement and participation.

As the "nature versus nurture" debates accelerated, Harding (1986), among others, asked a fundamentally different question. What is it about science and mathematics that women find unattractive? Is science inherently and fundamentally sexist? From a radical feminist perspective, Harding's thesis is that:

> [T]he epistemologies, metaphysics, ethics, and politics of the dominant forms of science are androcentric and mutually supportive; that despite the deeply ingrained Western cultural belief in science's intrinsic progressiveness, science today serves primarily regressive social tendencies; and that the social structure of science, many of its applications and technologies, its modes of defining research problems and designing experiments, its ways of constructing and conferring meanings are not only sexist but also racist, classist, and culturally coercive. (p. 9)

The empirical approach of modern science has both set the standard for scientific inquiry and separated scientific from nonscientific reasoning. Physics and chemistry in particular are held as the paradigms of scientific purity, abstractness, and certainty. The "lesser" sciences, including biology and anthropology as well as psychology, sociology, and education, were held to the standards of physics with the same expectations for computational confirmability and certainty.

In the 1950s, Quine challenged the assumptions of the scientific method. In his "Two Dogmas of Empiricism" he argued that the underlying assumptions of empiricism were "ill-founded":

> Modern empiricism has been conditioned in large part by two dogmas. One is the belief in some fundamental cleavage between truths which are *analytic*, or

> grounded in meanings independently of matters of fact, and truths which are *synthetic,* or grounded in fact. The other dogma is *reductionism:* the belief that each meaningful statement is equivalent to some logical construct upon terms which refer to immediate experience. (Quine, 1953, p. 20)

The basis for these dogmas of empiricism, he claimed, should be challenged and, if abandoned, the line between scientific knowledge and speculative knowledge would become less clear. These arguments will become pertinent to our later discussions on setting the logical foundations of inquiry in postmodern logics of process, systems, and meaning.

The underlying logic of the scientific method is fundamentally hegemonic, creating an imbalance in ways of knowing, emphasizing and elevating the status of scientific rationality over other forms of knowing. I will delineate the relationship between the logic of domination and the mathematization of reality, fundamental to the scientific paradigm below. Finally, I will show how the logic of domination inherent in modern science and mathematics has created an environment that is not only sexist and racist, but also damaging to our natural environment.

Underlying logic of modernism

According to Warren (1996), the logic of domination is a "structure of argumentation which leads to a justification of subordination" (p. 21). The logic of domination is inherent in the modern perspective and fundamentally distinguishes modernism from the premodern perspective. The logic of domination characterizes the modern era as the underlying meaning structure of the times. The logic of domination along with other aspects of the modern paradigm, namely value-hierarchical thinking and value dualism, would characterize modernism as an "oppressive conceptual framework":

> An *oppressive conceptual framework* is one that explains, justifies, and maintains relationships of domination and subordination....[T]here are three significant features of oppressive conceptual frameworks: (1) value-hierarchical thinking, i.e., "up-down" thinking which places higher value, status, or prestige on what is "up" rather than on what is "down": (2) value dualisms, i.e., disjunctive pairs in which the disjuncts are seen as oppositional (rather than complementary) and exclusive (rather than as inclusive), and which place higher value (status, prestige) on one disjunct rather than the other...and (3) logic of domination, i.e., a structure of argumentation which leads to a justification of subordination. (pp. 20–21)

To this list, Warren (1997) adds "power-over conceptions of power" and "conceptions of privilege, which serve to maintain and justify the dominance of those who are "up" over those who are "down"" (p. 20). She claims the central agenda for ecofeminism should be to expose oppressive conceptual frameworks and their underlying logic of domination.

As discussed in previous sections, fundamental to the rise of modern science was the mathematization of reality. Once nature was mathematized, the methods and inquiries of science were seen to provide the most objective and certain forms of knowledge. Tracing the historical emergence of the scientific paradigm from a feminist perspective, we can see the relationship between the mathematization of reality and the logic of domination. In her discursive treatment of the history of science, Griffin (1980) traces the separation of "man" from nature and the domination of man over nature and woman through the following themes:

> Separation of man from woman and nature
> Woman and nature to be overcome
> Taming the wild
> Civilizing the world
> Division of space and time
> Measurement as isolation and distinction
> Determination of the real through *his* senses and reason
> Rejection of emotion
> Control over nature and man's invulnerability
> Eliminating fear of uncertainty
> Certainty with computation
> Quantification of reality

It is clear how the logic of domination is intricately interwoven in both the scientific paradigm and the mathematization of reality. Merchant's (1980) classic book about the age-old identification of women and nature further supports the claim that modern science has as an underlying perspective of the logic of domination, the control of women/nature and the utilization of women's/nature's resources for the purposes of man. As described by Merchant (1980):

> The ancient identity of nature as a nurturing mother links woman's history with the history of the environment and ecological change. The female earth was central to the organic cosmology that was undermined by the Scientific Revolution and the rise of market-oriented culture in early modern Europe. (p. xx)

She critiques the influences competition, male aggression, and domination had on the environment and on women from feminist and ecological perspectives. By considering the implicit values of modern society, which include exploitation, control, rationality, and progress, she traces an alternative route to the relationship between modern science and the logic of domination foundational to it. The logic of domination and modernism, from her perspective, have been detrimental to women and our natural environment. The underlying logic of modernism, the logic of domination, has significantly impacted education and the curriculum.

The Curriculum and the Logic of Domination

The curriculum, based in the modern paradigm, similarly reflects and perpetuates the oppressive framework of value-hierarchical thinking, value dualisms, and the logic of domination. The curriculum hierarchy, as discussed above, reveals both a structure of mathematics and science as dominant over other studies and a value system that fragments thinking and values mathematics and science more than the humanities and arts. The purposes of education to prepare workers and perpetuate the mechanisms of society reveal a belief system that values technical control over feeling, thinking, and experience. Walkerdine (1988) argues that the role mathematics plays in society is part of an invisible, covert social control of rationality.

> It is my contention that the modern order is founded upon a rational, scientific, and calculating form of government, a government which claims to describe and control nature, according to natural laws. Thus mathematics can be understood as absolutely central to the production of this order.... Teaching children to read and write proved not to be enough to stamp out rebellion. The liberal order of choice and free will had to be created by inventing a natural childhood which could be produced and regulated in the most invisible of ways. Mathematics became reason. (pp. 211–212)

The modern curriculum reflects this perspective and touts competition over cooperation, individualization over community, and progress over process. These emphases are immediately traceable to our modernist roots and the underlying logic of domination of modernism. Even notions of time, by fragmenting the curricular day into definable slots of time and separating courses along disciplinary categories, is an aspect of the influence of modernism on the curriculum. As described by Hargreaves (1994):

> Time structures the work of teaching and is in turn structured through it. Time is therefore more than a minor organizational contingency, inhibiting or facilitating management's attempts to bring about change. Its definition and imposition form part of the very core of teachers' work and of the policies and perceptions of those who administer it. (p. 95)

The modern perspective has accomplished great things, but at what cost? Modern science, the development of technology, political systems including democracy and free enterprise, separation of church and state, and public education are all products of the modern perspective which, at the beginning of the 21st century are being challenged as detrimental to our individual and collective well-being and the future of our planet.

Modern Landscapes as Universals

The mathematization of reality and the ideas that numbers and numerical relationships can somehow be mapped onto the world has affected the development of the economies and political structures in existence today (see Capra, 1984, Merchant, 1980) as well as the educational landscapes. Educational paradigms are dialectically related to the larger social, political, and economic backdrop of our times and are affected by our perceptions of space and time. The production metaphor of schools so prevalent today is a direct consequence of our postindustrial society built upon the hypotheses of the mathematization of nature. The neoconservative rise is a deaththrow grasping for the underlying tenets of modernism from which the greatness of 20th-century America has been built. These attempts to reclaim our past while ignoring the problems and injustices of the modern mind-set are exactly the approaches of the neoconservative movement.

Despite efforts to grasp at the straws of our own past greatness, fundamental challenges have occurred in the sciences over the past century. These scientific revolutions, described in the next chapter, along with multiple social, economic, religious, and political influences, are marshaling in a postmodern era.

Postmodernism has been defined in a variety of ways, as a "diverse intellectual movement" (Fletcher, 2000, p. 28), "a reaction to, critique of, or departure from 'modernism,' to which the Enlightenment gave birth" (Schwandt, 2001, p. 120), "a symptom, not a fresh solution" (Latour, 1993, p. 46), and "an 'incredulity toward metanarratives,' those all-embracing foundations that we have relied on to make sense of the world" (The Internet Encyclopedia of Philosophy). More specifically:

> Postmodern theory has delegitimized what Lyotard has termed the Western "grand narratives" of liberalism, Marxism, philosophy and science....Postmodern thought suggests that the criteria which theories use to establish what is true or false, good or bad, are not universal and objective. They are, rather, in-

> ternal to the structures of the discourses themselves and thus historical and subject to change. (Weedon, 1997, p. 170)

Postmodernism is characterized as a rejection of foundations, universality, determinacy, and certainty. It is an "attitude" or "diagnosis" of our times.

> Postmodernism is an attitude toward the social world, more of a diagnosis than a theory. It opposes four central doctrines that form the core of the Enlightenment tradition: (1) the notion of a rational, autonomous subject; a self that has an essential human nature; (2) the notion of foundationalist epistemology (and foundationalist philosophy in general); (3) the notion of reason as a universal, a priori capacity of individuals; (4) the belief in social and moral progress through the rational application of social scientific theories to the arts and social institutions.... Postmodernism is also characterized by its distrust of and incredulity toward all "totalizing" discourses or metanarratives—those large—scale or grand theoretical frameworks that purportedly explain culture, society, human agency, and the like.... In place of these metaframeworks, postmodern theory endorses heterogeneity, difference, fragmentation, and indeterminacy. (Schwandt, 2001, p. 120)

The early 21st century is a transitory time. With conflict, discomfort, rapid change, and uncertainty come opportunity for change. Reaction to conflict can also result in retreat to a more comfortable past. It is not clear, in our 21st-century transitions, whether we will fundamentally transform our world and our relationships, social structures, and schools, or whether this will be remembered as a time of great darkness and regression. This book is written with the hope that we can transcend our modernist past and emerge with renewed connection, meaning, purpose, and value. The *Erwartungshorizonten* approach is intended to remind us of our past as we reach toward an uncertain future. The impact on and influence of schooling and education will be crucial in this postmodern period of transitions as we bring about our vision of the future.

CHAPTER THREE
Scientific Revolutions

> The term "revolution" as applied to science describes how better theories and evidence can sweep away established schools of thought. The real revolutions, however, take place inside our brains, when science helps us to understand the world anew.
> John Rennie, Editor, *Scientific American* 1999

As a former logician, philosopher, mathematics teacher, and computer teacher, I have been exploring the role of science in the development of Western intellectual and cultural history for a long time. I have been fascinated by the connections among scientific discoveries, approaches to inquiry, meaning and value, and philosophical investigations and movements related to the cultural heritage of the times. Especially in the modern era, the role of science has been key to defining who we are and how we approach our world.

This chapter details the changes occurring in the 20th century that have had an impact on society and fundamentally changed how we perceive science and knowledge. We are, arguably, within a period of scientific revolution where our own inquiries are leading us to see the world in a different way, to question that which we thought was most certain, and to challenge our own assumptions about the validity of and road to our most cherished ways of knowing. At the same time, the uncertainties created by these challenges have reminded us of our own humanity and reconnected us with our natural environment.

While this chapter may be hard to read for some who do not have a scientific background, the intent is to provide sufficient detail of the major scientific discoveries and movements of the 20th century so that the significance of the current scientific revolution may be explored. Because many of the postmodern metaphors are taken from the fields of quantum physics and theories of relativity, these ideas are explored in detail in this chapter. Similarly, many of the organic metaphors of postmodernism apply and misapply ideas from evolutionary theory and its many forms since before Darwin. For this reason, Darwin's theory of evolution will be distinguished from post-Darwinian and post-neo-Darwinian theories.

The relationship between philosophy and science has long been a close one. Dewey, Whitehead, Mead, as well as Aristotle, Descartes, and Leibniz saw no contradictions between philosophical inquiry and scientific exploration and addressed many of their investigations to the scientific discoveries and methods of their times. Mead (1934) asserted "the philosophy of a period is always an attempt to interpret its most secure knowledge" (Morris, 1967, p. ix). During the Modern era, science has dominated as the paragon of knowing, as discussed in Chapter Two. It is not surprising that science has similarly dominated philosophy during this period. Changes occurring in the sciences today are having significant impact on our culture and for this reason, it is important to understand the origins of sciences that are moving us beyond our modernist assumptions.

This chapter's exploration of the scientific revolutions of evolutionary biology, quantum physics, and theories of relativity complete the contextualization of the *Erwartungshorizonten* of the book. From the perspective of *Erwartungshorizonten* as described in these first three chapters, the logic and perspective for a dynamic curriculum will be set forth in Chapters Four through Seven, concluding with a vision of a dynamic curriculum in Chapters Eight and Nine.

Kuhn, in his books *The Structure of Scientific Revolutions* (1962) and *The Copernican Revolution* (1957), traces the anatomy of scientific revolutions. What distinguishes a scientific revolution from what Kuhn terms "normal science" is the paradigmatic shift necessitated by anomalies that can no longer be accounted for by the existing theory. The clockwork view of the world set in motion by the astronomy of Copernicus and culminating with Newtonian physics represented a significant break with previous scientific paradigms, Kuhn argues. The metaphor of the universe as a clock and the development of mathematics to model the universe (as we have discussed in Chapters One and Two), supported the development of the scientific method, with its underlying logic of domination and emphasis on measurement to ensure predictability, certainty, and control. The scientific method is the paragon of rational thought in the modern era.

Evolutionary biology, theories of relativity, and quantum physics represent, as a body, developments in science that have similarly impacted our perspectives of science as well as all of our social and intellectual paradigms. These developments can be considered the core of a currently existing revolution in the sciences impacting the whole of society and culture. These changes and developments began during the seventy-year period from the publication of Darwin's *Origins of the Species* in 1859 to the Bohr-Heisenberg-Einstein debates in the late 1920s. These findings in science initiated a general move away from specialization and a rejection of the fragmented approach of the scientific disciplines making room for relational, process, hermeneutical, and systems approaches developed after World War II. Quantum theory, theories of relativity, and evolutionary process theories collectively paved the way for underlying logics of relationship, meaning, and systems important for our postmodern, transitional times.

While understanding the nature of the scientific revolutions of the 20th century is important from our perspective of *Erwartungshorizonten,* understanding the *zeitgeist* that accompanies a fundamental shift in the way we view things is also important for exploring ideas about the curriculum. Breaking out of the endless circle of "innovation" that bounces back and forth between progressive and foundational movements in the curriculum will demand a similar revolution in our understanding of the curriculum and the social institution of schooling. The double-bind of scientific revolutions is that they represent both a shift in the *zeitgeist* of our times and generate new visions and metaphors that transform our ideas about schooling, the curriculum, and educative relationships.

Evolutionary Biology

Over the last fifty years, since World War II and the beginning of the Cold War, we have heard over and over how our students must be able to compete in order for democracy to be safe. The notions of competition, survival of the fittest, and performance standards continue to drive our curriculum. So fundamental to our beliefs about democracy that they go unchallenged are these ideas that in order to compete in the global marketplace we must "raise the bar" and "set high standards." Much of the rhetoric of the neoconservative agenda, including President George W.

Bush's education plan "No Child Left Behind," includes language and metaphors that can be traced back to Darwin's theory of evolution.

Developments in evolutionary biology were as significant as they were because they challenged the traditional perspectives of an orderly, stable, divine universe and presented a vision of developmental change, adaptation to the environment, and interactions among species with their environments. In direct opposition to the modern perspective that man is above nature, these scientific and philosophical movements reintegrated humans, humanity, and culture with nature. The Catholic Church reprimanded Teilhard de Chardin, a 20th-century scientist and Jesuit priest, for combining and addressing spirituality with evolutionary process and organic nature. For Chardin, contemplating the world in process rather than in fixed certainty was not a challenge to the authority of God, but in accordance with how God reveals himself to Man. "I am not speaking metaphorically," he wrote, "when I say that it is throughout the length and breadth and depth of the world in movement that man can attain the experience and vision of his God" (Chardin, 1965, p. 36). The tensions created by theories of evolutionary change originated in the debates between followers of Parmenides and Heraclitus in pre-Socratic times. Permanency and certainty are challenged by theories of evolution and change. The agenda for philosophy and science after the pre-Socratics, however, emphasized the former while process approaches were put on a side burner for almost 2500 years. These debates re-emerged in the 17th and 18th centuries as a rebellion against the mechanistic perspective and myopic vision of Newtonian physics.

Darwin's theory of evolution did not appear without precedent. His theory was the culmination of a movement that had explored in various fields the naturalistic progression of change. As described by Levins and Lewontin (1985):

> In 1859, when *Origin of the Species* was published, the evolutionary world view already permeated natural and social science. Evolutionary cosmology was founded in Kant's *Metaphysical Foundations of Natural Science* of 1786 and in Laplace's nebular hypothesis in 1796. Hutton's principle of uniformitarianism appeared in 1785 and became the dominant view in geology as a result of its centrality in Lyell's *Principles of Geology* of 1830. Evolutionary thermodynamics was begun by L.N. Sadi Carnot in 1824 and came to full development in 1851 in the work of William Thomson. In social science, Spencer's influence was enormous....English literature of the first half of the nineteenth century was thoroughly imbued with evolutionist ideology. (pp. 27–28)

In the intellectual traditions of the 19th century, ironically, biology was among the last to adopt an evolutionary perspective. The 18th- century Romantic poets, for example, addressed naturalistic, evolutionary, holistic perspectives in their objections to the mechanism of modern science and society long before similar objections were formalized in the sciences. In the field of biology, Lamarck and Saint-Hilaire made early attempts to account for the evolution of species at the beginning of the 19th century, before Darwin, as they observed that offspring seemed to inherit adaptive characteristics from their parents. Neither Lamarck and Saint-Hilaire nor Darwin, however, were able to describe the mechanism by which inherited traits and changes were passed on from generation to generation.

Darwin's theory delineated the principles for biological evolution as the Principle of Variation, the Principle of Heredity, and the Principle of Natural Selection.[1] While the Principle of Variation is unproblematic to most people and the mechanisms of the Principle of Heredity were basically unknown to Darwin and could not be explored until genetics and molecular biology developed in the early 20th century, the Principle of Natural Selection has been used in a variety of ways to support an underlying logic of domination.

Today much educational debate has occurred about the perceived conflict between the theory of evolution and the biblical version of the origins of man, including continued challenges to the teaching of the theory of evolution in schools. Many of these debates result from perceived implications of the Principle of Natural Selection in contrast with divine intervention. Not all interpretations of Darwin's theory challenge traditional perspectives of an orderly, perfect universe, however. Social Darwinism and Marxist political thought have applied implications of biological evolutionary theory to account for the evolution of political systems, economies, and social hierarchies as consequences of competition and survival of the fittest. Some even go so far as to imply there are "chosen" races more capable of adaptive change and possessing innate intelligence necessary for survival and competition. For example, in comparing how systems of thought are reflected in the economic systems of the day, Marx used Darwin's theory of evolution to explain both capitalism and socialism as logical consequences of modern society. The implied teleology of human evolution predicted, according to Marx, the final tension between these two economic and political systems. Herbert Spencer promoted Social Darwinism to justify differential social power as an indication of superiority and inherited traits based on race or ethnicity. In schools, we still employ a version of Social Darwinism when we use race, ethnicity, or socioeconomic status as predictors of success for entire classes of individuals. The principles of Variation, Heredity, and Natural Selection, leading to a survival of the fittest perspective, are evident in schools' organization and educator's explanations and expectations for student performance and "survival." They support a tendency to "blame the victims" whereby family backgrounds and individual traits are cited as reasons for success or failure rather than challenging the hegemony of our social institutions.

These social interpretations of Darwin emphasize conflict, competition, and survival with an assumption of limited resources. Implicit in these applications of Darwinian evolution are the ideas of teleological order and inherited destiny. These perspectives of evolutionary theory do not in themselves shift thinking away from Newtonian determinism. The emphasis in these theories on the mechanics of evolution resting in survival of the fittest and competition are consistent with the Newtonian perspective of the universe. The mechanism of Darwinism also removes our own guilt or responsibility for the cruel consequences of conditions as they evolve. Machines are set in motion; they are not run by intention, purpose, or value. This perspective of evolution as a natural process, devoid of spiritual intervention or human intentionality, mirrors rather than rejects the vision of a clockwork universe. As described by Rifkin and Perlas (1983):

> While it would be grossly unfair to suggest that Darwin knowingly borrowed concepts of industrial assembly, his theory of biological evolution did indeed reflect a similar method of production in nature. Each new species was seen as an assemblage of individual parts organized into new combination and arrangements and with additional improvements designed to increase both their complexity and their efficiency. (p. 98)

These interpretations of Darwin's theory reflected Darwin's own uncertainty about the mechanisms of evolution. While no hidden or invisible forces were invoked to account for species evolution, and while the mechanism for evolution seemed somehow tied to the varying environments in which species were found, how evolution occurred was not clear. Darwin could not account for natural selection and random mutation. Basing survival on successful competition similarly reflected a modernist mind-set and the emphasis on "bigger/more is better" and "might makes right."

This same lack of vision is still found in many "evolutionary" approaches to the curriculum. "Spiraled" curriculum, for example, calling for building on past successes and focused on content knowledge, still reflects a mechanistic perspective of growth grounded in this still very modern view of evolution. The defined path to take, the control of the learning environment to maximize survival potential, and the steps required to follow this path articulated in spiraled curricula reflect a mechanistic and Darwinian approach. Understanding evolution from a postmodern perspective requires re-examining the underlying logic and guiding questions of evolutionary theory. From this critical perspective, evolutionary theory can be recognized as a significant paradigm from which logics of process, relationship, and systems may emerge.

The discovery of and developments in gene theory combined with Darwin's theory of evolution resulted in neo-Darwinism in the early 20th century. The tenets of random mutation and natural selection combined with the chemical and physical properties of genes were used to demonstrate how evolution occurred. Neo-Darwinism, however, is still focused on the mechanics of evolution and is considered a modernist perspective emphasizing soul-less production and genetic superiority. Post-neo-Darwinism, (discussed in Chapter Five with the complexity sciences) challenges the idea that all variation occurs by chance and objects to the reductionistic perspective of examining and characterizing living processes as mathematical and chemical relationships alone. Post-neo-Darwinism tries to account for the "creative unfolding of life in forms of ever-increasing diversity and complexity as an inherent characteristic of all living systems" (Capra, 1996, p. 222).

Dennett (1995) argues that, despite its modernist interpretations and uses, Darwin's theory of evolution places evolutionary theory as a significant perspective from which postmodern ideas have emerged. Pre-Darwinian Newtonian mechanism often evoked an image of a supreme power, the master clockmaker, setting the universe in motion. Meaning and purpose in a mechanistic, soulless world requires such an invocation. One perspective of Darwinian thought was that it removed the clockmaker and introduced uncertainty and chance into the mechanism of change. Meaning, purpose, and answers to "why" and "what" (including "What is the meaning of it all?") seemed more at jeopardy with Darwin's theory of evolution. Thus, even a mechanistic interpretation of Darwin's theory challenges the clockwork universe paradigm. While order and stasis signified and were used to postulate a supreme power, Darwin's world of change and chance called into question the very need for or capabilities of an all-knowing, all-powerful god. The clockwork perspective of the universe both appealed to the notion of ideal creation and to predictable and certain relations among the component parts. These ideas were not consistent with evolutionary theory, species emerging and becoming extinct, chance mutation, and perpetual change. Existentialism and Romanticism were two intellectual movements in particular that emerged as responses to the apparent lack of meaning and purpose in a mechanistic world with no master helmsman. Dennett also argues that post-neo-Darwinian ideas reintroduced meaning, purpose, and value. These ideas (further explored in Chapter Five) are related to the underlying logic of systems.

Darwinian thinking was also significant because of its immediate impact on society. While the average person cared little about the developments in quantum physics as they unfolded in the early part of the 20th century, Darwin's theory of evolution was met with interest, resistance, criticism, and fear from the beginning. The Scopes Monkey Trial of 1925 is an example of the ferocity of debate and the divisiveness occurring as a result of the teaching of the theory of evolution in a Tennessee classroom.

In February 1925, a bill was passed in Tennessee that made it unlawful to teach in any public school "any theory that denies the story of the divine creation of man as taught in the Bible, and to teach instead that man has descended from a lower order of animals." As described by Knappman (1994):

> A few weeks later, in the little town of Dayton, a transplanted New Yorker with Darwinian views got into a debate at the local drugstore soda fountain with two Fundamentalist lawyers. However much they fought over evolution and whether mankind and monkeys were close relatives, they quickly agreed that a trial to test the law would do wonders for Dayton's commerce. The 24-year-old science teacher of the local high school, John Thomas Scopes, was recruited that very afternoon to be the legal guinea pig.... Using a state-approved textbook, Scopes taught a lesson on evolutionary theory on April 24 to his Rhea County High School science class. Arrested on May 7, Scopes was quickly indicted by the grand jury, setting the stage for what newspaper headline writers were already calling the "Monkey Trial." (p. 89)

Although later reversed on a technicality, initially Scopes was found guilty and fined $100 in a two-week trial that pitted Chicago attorney Clarence Darrow for the defense against the Fundamentalist rural lawyer and three-time presidential candidate William Jennings Bryan. The decision was reversed on a technicality. The trial itself was a news media sensation that reflected the interest in and controversy over the theory of evolution. Bryan proclaimed in a sermon to the people before the trial began that "the contest between evolution and Christianity is a duel to the death.... If evolution wins in Dayton, Christianity goes" (Knappman, 1994, p. 90).

I often show portions of the 1960 movie portrayal of the Scopes Trial, *Inherit the Wind,* to my classes as we discuss paradigms. The trial, as portrayed in the movie, seems to reflect this perspective of conflict between religion and science, faith-based reasoning and scientific rationality. The underlying logics of religious doctrine and scientific method are pitted against each other throughout the movie. Toward the end, however, we see how the Clarence Darrow character, played by Spencer Tracy, comes to understand the significance and dangers of pitting religious and scientific doctrines against one another. The movie ends not with the death of religion and the triumph of science, nor the replacement of a God-driven universe with an equally mechanistic universe with no Prime Mover, but with a hope for greater understanding and reconciliation. Maintaining the "divided" house of religion and science, we are destined, the movie suggests, to "inherit the wind." Looking beyond these divisions, we see the rising of the phoenix of postmodernism from the ashes of this conflict, not as a reconciliation of the scientific and religious paradigms, but as a transformation of both.

Teilhard de Chardin describes the transformation of perspectives to an organic, spirit-filled, changing universe as our ultimate and imminent spiritual quest.

> [T]he universe is being transformed and is maturing all around us.... Mankind is visibly passing through a crisis of growth.... The progress of the universe, in particular of the human universe, does not take place in competition with God, nor does it squander energies that we rightly owe to him.... The world can no more have two summits than a circumference can have two centers.... To desire the Parousia, all we have to do is to let the very heart of the earth, as we christianize it, beat within us. (Chardin, 1965, pp. 150–154)

Evolutionary theory is significant for the crisis of meaning and purpose introduced by ideas of change that are not crafted by a supreme being but seem to occur by chance. This crisis of meaning, a consequence of the underlying logic of domination and mechanistic perspective, is crucial to the current postmodern position of the early 21st century. Standing at the threshold of doubt, looking beyond the door of modernism, we are exploring how to rediscover meaning, purpose, and value. Spiritual, ontological, epistemological, and moral concerns discussed in relation to theories of evolution have contributed to our break from the mechanistic perspective. Evolutionary theory, as a scientific revolution, has opened the door to postmodernism.

Postmodernism, as a condition of late 20th and early 21st century Western society reflects the conflict of paradigms and underlying logics of modernism. However, it does not signify the arrival of the divine milieu described by Chardin. To continue the metaphor of postmodernism as the unlocking and opening of a door, we are not yet certain what is on the other side of the door. As our eyes adjust to the light, however, we are certain the door has opened to a new and different world. Other keys unlocking the door of postmodernism are quantum physics and theories of relativity. Quantum physics and theories of relativity have rocked the foundations of the scientific community and, while not as visible to the overall public, have challenged the modernist assumptions of technical rationality and scientific process. Each of these perspectives will be discussed below. Together with theories of evolution, these scientific revolutions represent significant challenges to the Newtonian clockwork universe and contribute to our current postmodern condition. Underlying logics of relationship, systems and meaning emerge from these paradigms to serve as the basis for a postmodern curriculum, the curriculum that will prepare our children for living on the other side of the door.

Einstein's Theories of Relativity

The Newtonian clockwork universe is described from an Archimedian perspective. Archimedes (287–212 BC) was a Greek philosopher, scientist, and mathematician. Among his many inventions was the lever, used to defend Alexandria from the Romans. He had boasted to King Ptolemy of Alexandria once about his invention—that if he had an immovable place to stand, he could move the world. The Newtonian worldview posits such a stance—the only way to understand the clockwork world of motions and forces is from this Archimedian perspective. Implicit in this perspective is the assumption of Absolute Space, Time, and Motion. Einstein challenged and confronted these absolutes.

Einstein published the specific theory of relativity in 1905 while working as a patent officer in Zurich, Switzerland. That year he published three significant papers including papers on quantum theory, statistical mechanics, and Brownian motion. It was in this third paper "On the Electrodynamics of Moving Bodies" that Einstein's specific theory of relativity was discussed.

Most significant about Einstein's theory is that it challenges Newtonian ideas about universal time and space. It wasn't that Newton didn't recognize relative relationships as we move about on earth, but his cosmological ideas rest on the notion of absolute space and time. As described by Gribbon (1995):

> Newton himself, although he realized that motion is relative as far as people moving about on the Earth, or birds flying through the air, or boats sailing on the sea are concerned, also thought that there must be some ultimate frame of reference against which all motion could be measured—a universal standard of rest. The idea of the ether fitted this picture, with all motion, measured relative to the ether. Newton also believed that there was an absolute standard of time, a kind of clock of God, which ticked away inexorably at the same rate for everybody. (p. 75)

The Michealson–Morley experiments in the 19th century to detect the ether, failed, however, calling into suspicion the postulated substance that could explain all motion. Einstein challenged these assumptions about space and time and rejected the hypothesized ether. He built his theories of relativity on electromagnetic field theory and theories about the speed of light. As described by Einstein in his 1949 essay on "The Theory of Relativity" (Einstein, 1956):

> The name "theory of relativity" is connected with the fact that motion from the point of view of possible experience always appears as the *relative* motion of one object with respect to another (e.g., of a car with respect to the ground, or the earth with respect to the sun and the fixed stars). Motion is never observable as "motion with respect to space" or, as it has been expressed, as "absolute motion." The "principle of relativity" in its widest sense is contained in the statement: The totality of physical phenomena is of such a character that it gives no basis for the introduction of the concept of "absolute motion"; or shorter but less precise: There is no absolute motion. (p. 39)

Another significant aspect of Einstein's theories of relativity concerned the relationship between space and time. Newton's ideas about time and space treated each separately. Motion was not considered by Newton to affect time and the space through which we moved was considered empty. The motion of mechanical devices, such as clocks, could be used to measure time, but time itself existed independently, according to Newton. Einstein's perspective was in direct opposition to Newton's. As quoted in Fritzsch (1994), Newton defined absolute space and time as follows:

> Absolute, true, and mathematical time, of itself, and from its own nature, flows equably without relation to anything external, and by another name is called duration.... Absolute space, in its own nature, without relation to anything external, remains always similar and immovable. (p. 13)

According to Newtonian theory, the homogeneity of space and time allowed a comparison of objects in space. The "absoluteness" of space and time included an absoluteness of comparison essential for mathematical reasoning. Two triangles are equivalent because they both have three sides of length six centimeters, for example. Their relative relationships to other figures on a page are irrelevant in determining their equivalency. Absolute space and time also entails infinite space and emptiness. Space is just "there," there is no need to consider the space in which an object is positioned in making a comparison with another object. We make these same assumptions in education. We don't need to consider the context of education, from a modernist perspective, in order to judge the efficiency of methods or individual understanding. Studies to examine the effectiveness of the use of manipulatives in mathematics or the superiority of whole language instruction in literacy development, for example, attempt to control for extraneous variables rather than exploring how context and individual differences may impact the effectiveness of any instructional strategy. Statistical methods often assume homogeneity rather than appreciate and seek out diversity. Aggregated data similarly ignores individual differences while assuming measures of "averages" give us meaningful information.

Einstein's theories of relativity challenge this kind of comparison of objects, irrelevant of contextual relationships. According to Einstein, time cannot be separated from space and there is not an assumption of the homogeneity of space. Applied to educational settings, the complexity of relationships and uniqueness of individual classrooms are as important as the individual teaching methods and student learning styles. Denying homogeneity of space, and recognizing the dynamic relationship of the evolving or unfolding occurrence of space-time, challenge modernist assumptions about relationships of causality and determinism. There can be no best method for teaching, and no universal assumptions about maximizing learning or efficiency of teaching. Context becomes a crucial construct for understanding learning differences.

Field theory further challenged the concept of Newtonian empty space. Objects in motion, according to theories of relativity, are in relative motion, that is, motion relative to something else. Fields of relationship define motion from this relativity perspective. Gravity, for example, was interpreted by Einstein as a magnet field rather than an invisible force as postulated by Newton. Objects interacting within a specific gravitational field do so relative to each other. Their reactions and interactions are systemic. Fish in the ocean, for example, continue to interact despite the swells and currents of ocean dynamics. Their relative interactions occur within the ocean's motion but are not caused by it. Similarly, gravity creates patterns of ways of interacting. As the swells in the ocean may cause entire schools of fish to bob and weave in unison in the water, so does gravity affect relative but not specific behavior. Field theory relies on the rejection of space as empty and considers space as an intricate field of interrelationships. As described by Wheatley (1994):

> If we were to observe fish, unaware of the medium of water in which they swim, we would probably look for explanations of their movements in terms of one fish influencing another. If one fish swam by, and we observed the second fish swerving a little, we might think that the first was exerting a force on the second. But if we observed all the fish deflecting in a regular pattern, we might begin to suspect that some other medium was influencing their movements. We could test for this medium, even if it were still invisible to us, by creating disturbances in it and noting the reactions of the fish. The space that is everywhere, from atoms to the sky, is more like this ocean, filled with fields that exert influence and bring matter into form. (p. 50)

In classrooms and schools, there are fields of interrelationship that, while noncausal and invisible, seem to permeate the context of our interactions. We observed how entire freshmen classes may take on a particular personality or how, walking down the hall, we get a sense of a school as a caring, warm place. These "invisible" fields are not causal forces but interconnected and complex networks of relationships that pervade the entire system.

If both space and time are relative to frames of reference and if space-time is inextricable within a specific gravitational field, the objective observer must be replaced by the participant observer representing a particular vantage point or frame of reference within the space-time continuum. There is no Archimedian vantage point for the relativity scientist! Space, time, and motion are inextricable and any measurement can only be made within a frame of reference of these three.

Einstein's general theory of relativity, developed and published in 1915, extended his specific theory of relativity by considering the impact of gravity, acceleration, and motion through curved space on the space-time continuum. Together his theories of relativity represent a significant break with Newtonian perspectives and challenge the certainty and mechanics of the clockwork universe. The significance of these ideas is not only that absolute space and time are abandoned, nor even that space-time is a continuum, but that the relativity of the frame of reference fundamentally challenges the possibility of objective observation or knowledge from an "outside" perspective. As described by Davies (1995):

> The major upshot of his new theory of relativity, then, was the prediction that time and space are not, as Newton had proclaimed, simply *there,* fixed once and for all in an absolute and universal way for all observers to share. Instead, they are in some sense *malleable,* able to stretch and shrink according to the observer's motion. (p. 53)

The agenda of modern science, and education following in its footsteps, is challenged as questions of objectivity, method, and certainty are entertained. The tree of science and certainty of method have been shaken and all the squirrels at the end of the branches, including education, are hanging on for dear life.

A third feature of Einstein's theories of relativity had to do with the implications of the relationship between energy and matter. Einstein's now famous equation, $E = mc^2$ (where c is the speed of light, m is mass, and E is the energy dynamic between mass and speed), similarly challenged the Newtonian and 19th-century ideas about the relationship between matter and energy. Related to the general theory of relativity, this equation suggests several consequences of the space–time continuum. Implied by this equation are that mass and energy are in a dynamic exchange. Velocity at or faster than the speed of light is impossible (without having infinite mass). Space-time is affected by velocity and gravity. Finally, time cannot be measured except from within a frame of reference. The lack of uniformity in the space-time continuum, the effects of velocity and gravity on time and motion, and the relativity implicit in a field of interaction fundamentally challenge the modernist agenda of science to find universal laws of nature in an orderly, timeless universe.

Thus, Einstein's theories of relativity have several basic tenets. First is the inextricable relationship between space and time, rejecting the Newtonian tenet of absolute space and time. Second is that motion is defined within a frame of reference, so there is no absolute motion, space, or time. Third is that the speed of light is a constant from which the relationship between energy and matter can be derived. Fourth is that gravity is a field that also affects the space-time continuum. The implications of these ideas set the stage for quantum physics, discussed below. Without quantum physics, however, it is unlikely that Einstein's theories of relativity would have had the impact on our thinking that they have. Theories of relativity lead to difficulties at the quantum level and, together with quantum theory, open the door of postmodernism a bit more.

> Relativity theory itself, while having important consequences for the way that some physics is done, is not likely to lead to a new world view.... [R]elativity theory itself is about the physics of high velocities and very great distances. It

> plays itself out on a cosmological scale and has virtually no application in our everyday, earthbound world.... Quantum physics is different. Being the physics of that tiny microworld within the atom, it describes the inner workings of everything we see and, at least physically, are. (Zohar, 1990, pp. 20–21)

In education, challenges to the Archimedian stance and foundations of modern science have serious implications for many of our most sacred truths. Standardized testing, intelligence, aptitude, knowledge, and truth start to become suspect when relativity sciences are recognized. With quantum physics, any hope of meaningful measures and truth is dashed.

Quantum Physics

Building upon the discoveries of Faraday in electromagnetic field theory, Maxwell in the latter part of the 19th century explored and formalized key relationships between thermodynamics, electricity, and light. Properties of light were explored as electromagnetic fields. Maxwell's equations seemed to prove the speed of light was constant. He introduced the constant c for the speed of light and experimentally measured the speed of light to be approximately 300,000 km/sec^2. Treating space as electromagnetic fields, light as a wave, and the speed of light as a constant were essential ingredients for Einstein's theories of relativity as described above.

While these discoveries were essential for the theories of relativity and cosmology, physicists at the same time were working on understanding electromagnetic relationships at subatomic levels. With advances in chemistry and biology, the agenda of modern physics, chemistry, and biology in the late 19th century was to determine the basic building blocks of nature. Consistent with the Newtonian clockwork perception of the universe, the underlying premise was, if we can determine the nature of the basic elements of nature, we can understand how nature is "put together," how it works, how to take it apart, and how to put it back together.

Planck's discoveries in electromagnetic radiation at the turn of the century had determined that electromagnetic energy was emitted in packets of particular sizes, named quanta by Planck. As described by Capra (1984):

> The situation seemed hopelessly paradoxical until it was realized that the terms "particle" and "wave" refer to classical concepts which are not fully adequate to describe atomic phenomena. An electron is neither a particle nor a wave, but

> it may show particle-like aspects in some situations and wave-like aspects in others. (p. 79)

Einstein's famous equation, $E = mc^2$, however, was particularly troublesome since the relationship between matter and waves as bundles of energy became blurred. As further described by Capra (1984):

> The discovery that mass is a form of energy has had a profound influence on our picture of matter and had forced us to modify our concept of a particle in an essential way. In modern physics, mass is no longer associated with a material substance, and hence particles are not seen as consisting of any basic "stuff," but as bundles of energy. Energy, however, is associated with activity, with processes, and this implies that the nature of subatomic particles is intrinsically dynamic. (p. 90)

The group of physicists working on the problems of quantum physics in the first three decades of the 20th century included Planck, Einstein, Bohr, DeBroglie, Schrodinger, Pauli, Heisenberg, and Dirac. Two important findings of their debates of the late 1920s were Bohr's Principle of Complementarity and the Heisenberg Uncertainty Principle.

The Principle of Complementarity suggests the problem of solving the dual nature of quanta as more fundamentally matter or energy rests on recognizing the complementary relationship between these two states. Quanta at their most basic level are both matter and energy. As described by Wheatley (1994):

> [E]lementary matter is inherently two-faced. It possesses two very different manifestations. Matter can be particles, localized points in space; or it can be waves, energy dispersed over a finite volume. Matter's total identity (known as a wave packet) includes potentialities for both forms—particles and waves. This is the Principle of Complementarity. (p. 35)

The significance of the dual nature of the most basic building blocks of nature challenges the belief in underlying matter or a single reality. It also responds to the dualisms inherent in the matter-energy debates with a "middle-ground" answer. Rather than choosing sides in the debate, the solution seems to be that subatomic relationships are more complex than our previous theories had postulated. This response, also evident in the Heisenberg Uncertainty Principle, challenges traditional Cartesian either-or logic. The Law of Contradiction (P and not-P cannot both be true) and the Law of the Excluded Middle (either P or not-P must be true) in mathematical logic don't seem to apply to quanta, as revealed by the Uncertainty Principle and Principle of Complementarity!

The Heisenberg Uncertainty Principle is related to the Principle of Complementarity as a response to the paradoxes of quantum physics. Like the Complementarity Principle, the Uncertainty Principle rejects the dualisms implicit in Cartesian thinking. While the basic properties of quanta are both wave and matter as described by the Principle of Complementarity, the Uncertainty Principle suggested that we can never observe both states at the same time. As described by Zohar (1990):

> And while neither "state" is complete in itself, and both are necessary to give us a complete picture of reality, it turns out that we can never focus on both at once.... According to the Uncertainty Principle, the wave and particle descriptions of being preclude one another. While *both* are necessary to get a full grasp of what being is, only *one* is available at any given time. (p. 26)

The Complementarity Principle together with Heisenberg's Uncertainty Principle is similar to the Gestalt idea that many of our perceptions reveal the dual nature of figures. As with the Janus face or the two-dimensional stick representation of the cube, we can see two separate images at different times, the old woman or the young girl, for example, but never both at once. The choice to examine one image, say that of the girl, brings forth that vision just as the decision to measure quanta as waves seems to make them behave as waves.

Together, these two principles challenge the assumption of an underlying reality or being. It seems being is in process, interacting not only within its own dual nature but interacting with us as we attempt to observe its properties. These fundamental principles of quantum physics support a process theoretic emphasis on relationship, and "being" as process and interaction. This becomes clearer when measurement is explained further.

If the scientist chooses to measure quanta as particles, they behave as particles; if, however, the scientist decides to measure the wave pattern relationships of quanta, they behave like waves or energy. The interactive relationship between the act of observation and what is observed suggests an intimate relationship between what we observe and our expectations of observation, including the kinds of measures we use. Thus, in direct opposition to the realist's notion of the objective observer who observes the underlying truth of reality by employing appropriately objective measures and procedures, the Heisenberg Uncertainty Principle suggests our interactions with and choice of instruments we use to measure in some way brings forth the kind of information we observe. If we set up our experiment to observe quanta as waves, they will behave like waves. Similarly, if we try to examine their material qualities, they behave like matter. That the choice of measurement seems to be anticipated by the system is a problem from a realist perspective. As described by Prigogine and Stengers (1984):

> That quantum mechanics obliges us to speak less absolutely about the localization of an object implies, as Niels Bohr often emphasized, that we must give up the realism of classical physics. For Bohr, Planck's constant defines the interaction between a quantum system and the measurement device as nondecomposable. It is only to the quantum phenomenon as a whole, including the measurement interaction, that we can ascribe numerical values. All description thus implies a choice of the measurement device, a choice of the question asked. In this sense, the answer, the result of the measurement, does not give us access to a given reality. We have to decide which measurement we are going to perform and which question our experiments will ask the system. Thus there is an irreducible multiplicity of representations for a system, each connected with a determined set of operators. (pp. 224–225)

The implications in education are tremendous. What if Binet had developed the I.Q. test emphasizing mechanical and artistic skill rather than analytic and verbal skills? Certainly many of us who have been privileged in school because we were "smart" would find ourselves in remedial classes! Where would mathematics and language arts be in this curriculum in a society that measured intelligence by artistic talent? What would our weapons of war be like? How, ultimately, would our mathematics and language skills have evolved? How is what we teach influenced by what we can measure? And how is what we can measure an indication of what we value?

That our choices of what to measure in some fundamental way bring forth what we "see," challenges the assumed relationship between mathematics and nature considered as valid during the Modern Era. Measurement, rather than being the arbitrator and gauge of reality, becomes an integral part of defining what is observed through our own interactions. Rather than a map to the universe, measurement brings forth a universe. A change in measurement changes our universe! This change is not just a change in our perspective, but seems to reflect a fundamental change in what we are measuring.

Similarly, the Principle of Complementarity, that "there is an irreducible multiplicity of representations for a system," suggests:

> No single theoretical language articulating the variables to which a well-defined value can be attributed can exhaust the physical content of a system. Various possible languages and points of view about the system may be complementary. (Prigogine & Stengers, 1984, p. 225)

While in schools we may do a better job with the notion that multiple measures are necessary to get a true picture of what a student knows and can do, we still believe our measures are measuring something. If we are trying to access a learning disability, for example, we may use a variety of performance tasks as well as standardized measures to determine the extent of the disability. We nevertheless believe we are getting a "truer" or "clearer" picture of the problem rather than recognize that the problem exists, not because the child had a traumatic brain injury but because we have an idea of learning disabilities that we are trying to measure. Our measurement and our desire to measure brings into existence many ways of seeing our children rather than revealing underlying truths or realities about them.

The consequences of settling the question of the dual nature of subatomic quanta by recognizing both aspects of quanta were significant. As described by Capra (1984):

> The resolution of the particle/wave paradox forced physicists to accept an aspect of reality that called into question the very foundation of the mechanistic world view—the concept of the reality of matter. At the subatomic level, matter does not exist with certainty at definite places, but rather shows "tendencies to exist," and atomic events do not occur with certainty at definite times and in definite ways, but rather show "tendencies to occur." (p. 80)

By the end of the 1920s, physicists accepted that the basic units of matter were both energy and matter; that the basic building blocks of our world were not so many subatomic blocks of matter, but rather probability patterns of energy with the dual character of waves and matter, intricately connected in webs of relationship. Finally, it was agreed, motion, at the subatomic level, does not occur in smooth progression but in jumps or leaps as discovered by Planck early on when he first defined quanta.

Quantum physics is thus the study of relationships and provides us with a perspective of the universe as *participatory rather than abstract*, *emergent rather than determinate*, and *discontinuous rather than continuous*. As described by Zohar (1990):

> Perhaps more than anything else, quantum physics promises to transform our notions of relationship. Both the concept of being as an indeterminate wave/particle dualism and a concept of movement that rests on virtual transitions presage a revolution in our perception of how things relate. Things and events once conceived of as separate, parted in both space and time, are seen by the quantum theorist as so integrally linked that their bond mocks the reality of both space and time. They behave, instead, as multiple aspects of some larger

> whole, their "individual" existences deriving both their definition and their meaning from that whole. (p. 34)

Another consequence of quantum physics is synchronicity or instantaneous nonlocality, as first demonstrated by Einstein. Experiments with quanta suggested there is a subatomic network of communications that occur among distant and seemingly unconnected quanta. These communications seem to occur instantaneously and asynchronicitously. As we enter a school for the first time, this same pattern of relationships and network of communications seem to allow us to get a "feel" for the school. There are no interactions or direct experiences to explain the feelings we get when we enter the school, yet the "sense of the place" seems to be as apparent as the children's artwork decorating the halls. Current brain research also suggests the brain may use synchronicitous communications, that memories and ideas live in many parts of the brain, and that the "simultaneous shout" of neural firings is what allows rapid processing as well as recovery after many brain trauma incidents.

The notion that there are causes and relationships that are not direct does not make sense in nor can it be accommodated by Newtonian physics. Especially problematic for the Newtonian perspective is that these relationships or networks may be networks of communications, meanings, or feelings! Patterns of relationship and order seem to provide a stability that, from a Newtonian perspective, should not occur without causal connections and relationships among "things." Like so many billiard balls on a pool table, connections across space and time must occur, according to the Newtonian perspective, only as the balls are set into motion and a string of cause-and-effect collisions relate the balls to one another, can the balls respond to one another.

Notions of synchronicity and instantaneous nonlocality also point back to theories of relativity and the idea that simultaneity is itself relative to a given frame of reference. Some have applied these notions to Jungian psychology with the hopes of integrating quantum theory and psychology. As described by Peat (1987):

> Synchronicities give us a glimpse beyond our conventional notions of time and causality into the immense patterns of nature, the underlying dance which connects all things and the mirror which is suspended between inner and outer universes. With synchronicity as our starting point, it becomes possible to begin the construction of a bridge that spans the worlds of mind and matter, physics and psyche. (p. 2)

Energy and field relationships seem to provide a way of seeing the world as a unified whole rather than pieces bumping into each other, forces causing reactions, and the dual nature of reality as comprised of minds in one realm, and matter in another. The relational and probabilistic universe portrayed by quantum physics is more consistent with a systems perspective of interconnectivity and mutuality and cannot be reconciled within the Newtonian perspective. For this reason, many theoreticians have used quantum physics both experimentally and metaphorically as the underlying science of relationship. The intricacies of interrelationship found at the quantum level transcend the nature of the perceived object at the macroscopic level of tables, chairs, and people, to suggest the relevance of field relationships which include synchronicity and non-causal interconnections and relationship as transcending and primary to space and time.

The origins and developments in systems theory, while emerging primarily in the fields of biology and thermodynamics rather than quantum physics, are essential for understanding the new sciences of chaos and complexity discussed in Chapter Five. The relevance of these scientific revolutions for underlying theories of relationship, meaning, and systems are explored more fully in the middle section of this book. Their relationship to the curriculum is further elaborated in the concluding chapters in the final section of the book.

Summary

The first three chapters have provided the perspective of the *Erwartungshorizonten* of the late 20th century. The impact of Newtonian science and the underlying logic of domination of modernism are still felt in our schools today. Scientific rationality; beliefs (and confusions) about the validity of measurement and the objectivity of reality, a curriculum that emphasizes mathematics and science over other subject matter; and the perceived crisis that the "American way" is in jeopardy if our children cannot "successfully compete" in a global marketplace, are all embedded in the modernist mind-set.

The first three chapters have also provided the origins for our developing theories from which a dynamic, postmodern curriculum may evolve. Technique, the next section of the book, will offer the approaches of complex adaptive systems theories, logic as inquiry, and logic of meaning based on the perspectives of process, relationship, language, synchronicity, relativity, and evolution developed in the first three chapters. The logics of process, systems, and meaning form the framework of our technique.

Notes

1 The Principle of Variation: Individuals within a species vary in physiology, morphology, and behavior. Varied characteristics occur through chance mutation. The Principle of Heredity: Offspring resemble their parents on average more than they resemble unrelated individuals. The Principle of Natural Selection: Different variants leave different numbers of offspring. (See Levins & Lewontin, 1985, p. 32.)

2 Maxwell measured the speed of light by experimentally measuring the speed of electrical impulses in wires. Michelson, in the 1870s actually developed experiments using mirrors, to measure the speed of light directly. See Davis (1995) or Gribbin (1995) for historical accounts of these experiments and findings.

PART II: TECHNIQUE

Holographic Images

CHAPTER FOUR
The Logic of Relationship

The modernist approach to public education in the West has emphasized that schools and the process of schooling are important aspects of our democratic society. Building the capacity for change, meeting the needs of a changing society, and preparing new generations for a technological future are goals, from this perspective, that must be realized in order for schools and public education to survive. Without a strong public education system and the ability to compete in the global market economy, the neoconservatives would have us believe that our very way of life is threatened.

Communities and schools have become battlegrounds around the standards debates, testing, international comparisons, limited resources, future viability of the Western way of life, and traditional values and concerns. Children, in increasing numbers, feel unsafe and unwelcome in their own schools. Physical and psychological safety are threatened as schools become less and less personal and emphasis is placed on the aggregate rather than the individual needs of learners. Learning and meaning are not yet the experience of many of our students as more and more of our children become alienated from school, their teachers, and each other.

Significant change, especially in light of rapid changes in society, will require more drastic measures than reforming existing structures. New metaphors that reject the factory model of schooling and the underlying logic of domination of modernism must be explored to overcome traditional ideas about how to educate our children.

While the underlying logics of process and relationship (Dewey, 1938; Whitehead, 1929), systems thinking (Angyal, 1941), and meaning (Wittgenstein, 1953) have been around for a long time, school reform efforts have not been approached from these perspectives nor have they had a significant impact on social structures at every level of society. Understanding and building capacity for schools as learning organizations may infuse schools with the spirit of adaptive change and recreate the heart of schooling. The logics of relationship, systems, and meaning synergistically contribute to a perspective from which the social system of schools may evolve as learning organizations.

Key features of a relational logic, discussed in this chapter, include a focus on process, logic as inquiry, and learning as a dynamic progression. Transformational learning, in particular, requires the development of critical reflection and the dialectic among theory and action. Whitehead's process philosophy and Dewey's approach to inquiry form the basis of this chapter to develop a perspective of a relational logic. The mechanistic approach to schools is challenged by a more organic approach of relational thinking. Related to Native American and Asian cultures as well as to quantum physics and New Science, the cosmic dance of relationship will be seen to interrelate change, unity, holism, chaos, order, emptiness, form, symmetries, space, and time.

While relational logic provides naturalistic rather than a mechanical way of understanding change, systems logic, discussed in Chapter Five, recognizes the importance of relationship from another perspective. Systems logic focuses on the notions that change cannot occur piecemeal and patterns of organization and emergence are central aspects of organic growth and change. Synergy or complexity, including the Gestalt idea that the whole is greater than the sum of its parts, captures the organizational dynamics of complex systems. Holistic, comprehensive, connected change must be approached from a systems perspective. Infusing social organizations with the flexibility and means of adaptation goes beyond the adaptive capabilities of the individuals comprising the organization. Chapter Five explores complexity theory and ideas about the relationship between learning and adaptation from a systems perspective.

Chapter Six explores the logic of meaning. Ideas about language, meaning, and logic as truth or structure have been challenged in the 20th century. The implication for these challenges is that what we traditionally have valued or taken as demonstrations of learning are not meaningful. Wittgenstein's logic of meaning as explored in his later works provides the basis for a postmodern approach to education and a different perspective on meaning, truth, and value.

Finally, Chapter Seven explores learning organization literature as applied to business and industry. Key features of learning organizations as they relate to process, systems, and meaning will be highlighted as strategies for revisioning schools as learning organizations are explored.

Together, these four chapters comprise the perspectives and visions necessary for challenging our own modernist tendencies. These are the lenses through which we will focus beyond the threshold of modernism where we now stand. The final section of the book, which includes Chapters Eight and Nine, will explore curriculum, instruction, and learning, informed by the logics of relationship, systems, and meanings, and within the dynamics of learning organizations.

Holism and Relationship

The arrogance of Western thinking has excluded considerations of non-Western cultures as having anything of significance to add to our own superior and advanced ideas, including our ways of knowing. The "primitives" or "ancients" have had "insights" or "visions" that are now valued not as challenges to our own culture but because of our own changing ideas about "reality," knowing, and being. Change, unity, holism, symmetry, form, and the view of a living, connected, moral universe in which we are only a small part, are ideas that Western traditions have tried hard to accommodate at the turn of this third millennium when previously these perspectives were rejected as quaint, unscientific, or ancient. Non-western, indigenous cultures have embraced these perspectives while we in the West have struggled in recent years to accommodate the consequences of our own modernist assumptions. Rather than recognizing the significance and wisdom of these non-western perspectives, however, we still filter them through our own Western culture. A more natural approach to the environment is interpreted by our society as legislation to fine those who pollute too much. For-profit and non-profit agencies are formed, to which we can contribute money, so that we can provide "safe zones" for endangered species. Advancements in science allow us to take control of the reproductive processes of animals so we can control their numbers. Genetic engineering allows us to have better beef to eat, a bigger yield on crops, and lower prices at the market. Fewer farmers are needed to produce higher yields, yet those who chose to make a living working the land find they cannot do so without becoming a part of huge farming enterprises. Our "advanced culture," which now has the luxury to discover the importance of our relationship with nature, has appropriated it from other cultures, twisted it to fit our modernist paradigm, and adopted it as yet another way of accomplishing technical control.

While the holism and connectedness of an underlying logic of relationship had been a part of our Western perspective from the beginning, for the most part, it has existed only as a challenge to modernism in the fringes of poetry, architecture, the arts, and the philosophical movements of process philosophy, existentialism, and pragmatism. Process philosophy, existentialism, and pragmatism represent concerted efforts to reject modernist assumptions about meaning, knowing, and being, yet these intellectual traditions themselves have been marginalized as have the romantic poets and other "fringe" thinkers. The production metaphor of schooling is so entrenched in the modernist mind-set that these non-scientific, intellectual, and cultural traditions that have been evolving for more than two hundred years have had little impact on how we educate our children.

My teaching radically changed when I stopped thinking of students, learning, knowledge, teaching, and schooling as "things" from a production perspective. Even ecological or growth metaphors didn't really work for me until I started thinking of students, learning, and schooling as *relationships* and *contextual*. This change in what I believe to be most fundamental, namely that students are complexes of relationships rather than things, living within individual and social contexts, has completely affected what I feel is important in my classroom, how I approach instruction and think about teaching, and how I view assessment.

A shift in thinking about "things" to thinking about "relationships" and recognizing the impact of interconnectedness and context are more than a change in perspective, however. These shifts in thinking represent a fundamentally different way of seeing and approaching the world and all that is in it. Exploring relationships rather than examining things has implications for every facet of life. The logic of relationship has its own demands for value, purpose, and method. Examining and shifting our ideas about the purpose of schooling with an underlying logic of relationship has the potential to infuse the curriculum with meaning and purpose, rekindling the spirit and heart of learning.

This chapter will address the logic of relationship found in Whitehead's cosmology and Dewey's pragmatism. Native American and Eastern perspectives remind us that the logic of relationship is grounded in holism and a perspective of our relationship with nature that has been the norm rather than the exception in most cultures. The logic of relationship, while traceable to process traditions in Western thought, has a rich tradition in these other cultures. The relationships and connections among Western process thinking, Native American wholeness, and Eastern interconnectedness will be discussed as the basis for an underlying logic of relationship.

Origins of Western Process Thinking

Process philosophy marks a significant shift from traditional philosophical perspectives. The focus of modern philosophy has been on epistemological questions of knowing and certainty. Whether we ultimately only know sense data as our experiences of the outside world (Idealists) or whether through our methods we come to certain and secure knowledge (Rationalists and Empiricists), the focus of Cartesian-based philosophies has been on knowing "things." Process philosophy focuses on ontology and questions of relationship and process. The shift in focus from "things" to "relationships" or from "products" to "processes" is not purely postmodern, however. Process philosophy has its roots in ancient Western philosophical perspectives.

In his opposition to Parmenides, the preSocratic Greek philosopher, Heraclitus, claimed the essence of all being was becoming. Rather than perceiving all of nature composed of discrete, separate entities, Heraclitus initiated a tradition whereby process, change, and relationship were perceived to be fundamental to all that is. Zeno's paradoxes were intended to counteract the perspective of Heraclitus and support the claims of Parmenides that our sensory experiences could be fooled and that only through rational thought could true knowledge be obtained. Our experience of motion is an illusion, Parmenides argued, since fundamental reality was permanent and unchanging. The arrow that appears to be in motion, at any particular time, is at rest. The motion is an illusion since it is impossible that the arrow traveling through space is really the infinite sum of arrow-moments at rest. But how, if motion is an illusion, are we to understand change, including spatial change as motion through space, as we experience it?

Plato's response to these problems was to emphasize that truth and reality are not of the world of experience, but are abstract, pure forms, unchanging and knowable not through empirical means but through reason. Mathematics is the single most important study for developing reason, according to Plato, and philosophy is the application of reason to unveil the truth.

While Plato construes being as undifferentiated, abstract form, Aristotle emphasizes and initiates the analytic tradition. Rather than one truth, one single (abstract) reality, Aristotle compartmentalizes being and knowledge. Knowledge is gained through empirical studies of reality. Among the empirical traditions, scientific study is the most important for Aristotle.

Atomism was also prevalent in pre-modern thinking. All being, according to Democritus and Epicurus, was comprised of basic units or building blocks. Similar to Aristotle in its analytic approach, atomism focuses on examining the whole of any system or structure by breaking down the structure to its most elemental parts. Atomistic approaches are therefore empirical and analytic.

The premoderns recognized the difficulties of accounting for change, identity, and growth. Their ontological theories included and were intimately related to their ethics and view of moral living. The truth and the good were inextricable from their ontological perspectives. Whether the goal was individual living for the good or maximizing the good in society, there was a strong tie in these premodern perspectives among the study of being, ethics, and morality.

Distinct from its premodern roots, the modern perspective ignores the reason these theories of being were advanced in the first place. Although the question of the ultimate reality is an important philosophical question in its own right, modernism forgot to ask "Why?" Why do we care what is essential to all being? Why is it important to have a theory of being? Why is knowledge important? During the Modern Era, as described in Chapters One and Two, these questions were answered in a way that reflects the underlying logic of domination and focuses on scientific rationality. In a break with the theism of the Middle Ages and the morality of ancient wisdom, science became a new God, which empowered man[1] to "tease" out nature's secrets and control his destiny.

> The fate of many epistemological modes was sealed by Galileo in the moment when he founded modern natural science. In order to understand the book of the universe, Galileo predicated, we must learn how to read the language in which it is composed, and the basis for interpreting the world of nature consists of "figures, numbers and movements, but"—and this is a turning-point in his scientific philosophy—"not smell, taste, nor sounds, which I do not believe are anything more than names outside the living animal." Galileo's expulsion of the domains of the auditory, the olfactory, and the sense of taste from the realm of scientific inquiry corresponds to a critical phase in human cognitive development as hypothesized by Vico—namely, the point when humans no longer depend on their physical senses as the matrix for thinking. In this manner, Galileo is very much in consonance with Descartes, who selected as his inter-

> pretive materials quantifiable data of which we "could be sure," unlike our physical senses, which would inevitably weaken our ability to produce a strong and lucent, linear form of thinking. (Colilli, 1997, pp. 50–51)

As discussed in Chapter One, the modern mind-set was born during the Renaissance or "reawakening" of intellectual thought made possible in part because of the dissemination of translations of ancient texts including the writings of the Greek thinkers. Renaissance thinkers were in direct opposition to the authority of the church and many, Galileo for one, were persecuted for or required to renounce their scientific theories. The tension between religious dogma and the emerging sciences of the Renaissance resulted in the abandonment of all scholastic thinking. As Whitehead (1925/1967) argues:

> Science has never shaken off the impress of its origin in the historical revolt of the later Renaissance. It has remained predominantly an anti-rationalistic[2] movement, based upon naive faith. What reasoning it has wanted, has been borrowed from mathematics which is a surviving relic of Greek rationalism, following the deductive method. Science repudiates philosophy. In other words, it has never cared to justify its faith or to explain its meanings. (p. 16)

The over-emphasis we have placed on science has provided fertile ground for the growth of science, while blinding us to the fact that the faith we have placed in science, the way we have deified scientific thinking and knowledge, has gone unexamined during the Modern Era.

> The faith in the order of nature which has made possible the growth of science is a particular example of a deeper faith. This faith cannot be justified by any inductive generalisation (sic). (Whitehead, 1925/1967, p. 18)

The logic of domination and the mathematization of nature significantly narrowed the traditional philosophical approaches of the ancients as modern society lost site of the ethical and moral dimensions of theories about being and uncritically accepted the tenets of modern science.

Whitehead's process philosophy, most thoroughly espoused in his 1929 book *Process and Reality,* is an attempt to recapture the holism of the ancients while accounting for the developments of quantum physics, evolutionary biology, and theories of relativity. In the 20th century, process philosophy was thus reborn through Whitehead as an attempt to address and overcome the fragmentation and isolation resulting from the unquestioning acceptance and predominance of scientific rationality and the logic of domination.

> [P]rocess philosophers utilize the data and theories of contemporary science to argue for the replacement of the mechanistic by a revised teleological paradigm.... [T]his contemporary form of teleological explanation tends towards pluralism, stressing finite, interacting centers of purposive activity.... It is in these various senses that process philosophy represents simultaneously a completely contemporary interpretation of nature, as well as a recurrence of a modernized variation of ancient Greek cosmology. (Lucas, 1983, pp. 9–10)

Whitehead is considered the originator of the modern process philosophy movement. Relevant aspects of his process philosophy are presented below.

Whitehead's process philosophy

In the early part of the 20th century, Whitehead's emphasis on process was used to develop a philosophy he had hoped would accommodate theories of relativity, evolutionary biology, and quantum physics. Rather than accepting the Newtonian trio of space, time, and matter as primitive, Whitehead's process philosophy focused on events as ultimate being while space, time, and matter were considered derivative concepts or abstractions.

> "Actual entities"—also termed "actual occasions"—are the final real things of which the world is made up.... The final facts are, all alike, actual entities; and these actual entities *are drops of experience, complex and interdependent.* (Whitehead, 1929/1978, pp. 27–28, emphasis added)

Key components of his process philosophy include the ideas that (1) events, as occasions of experience, are the fundamental reality, (2) a systems perspective allows for a more holistic and naturalistic understanding of relationship, and (3) patterns of organization are important for understanding how systems work. Understanding process and relationship should be the focus of inquiry, according to Whitehead. The significance of this shift is seen when we imagine what exploring student learning would be like if, instead of using standardized tests to determine what students have learned, that is, instead of considering the products of education, we explored the processes and relationships nurtured. What is this child's relationship with reading? How has he or she come to love mathematics? What are the patterns of experiences that have evolved for these children? What are their questions or concerns and how may we help them make sense of their world?

Underlying process is an understanding of "being" in relationship, through change. Understanding being not as Newtonian "things" or basics, but as complex processes has been a problem for process philosophers. What are the basic entities of reality if they are perceived not as things but as relationships or experiences? There are three ways to understand "being" as in process: to identify being with becoming, to suggest being is produced by its own becomings, or to recognize that being, in process, is constituted by its own history of becomings. Each of these approaches has been tried by process philosophers and will be discussed below.

Considering being-in-process as a collection of "becomings" addresses one aspect of the question "How is that which is most real or fundamental always in process and relationship?" This suggests some underlying structure to being-in-process that is the "real thing" throughout these changing and evolving internal and external relationships. This reproduces Zeno's paradoxes of time, raising the question: how can being be made up of infinite time-slices of becoming?

Another approach to understanding being is to consider being as the result of or produced by becoming. From this perspective, we are the product of our "becomings" (which, by different interpretations may be experiences, environments, or physical entities). This perspective of being has difficulty accounting for creativity and free will and falls to the various problems inherent in Newtonian determinism.

Whitehead viewed being as "constituted" by becoming. Process, according to Whitehead, therefore, is the becoming of experience. This view of process is key to a logic of relationship with an emphasis on experience and interrelationship. Events or occurrences as relational occurrences of experience are fundamental, for Whitehead. Being constituted by becoming, entails process. Being, as process, implicates experience, change, motion, and relationship. In particular, Whitehead maintains, being is "a dynamic process of creative advance that will never end" (Gragg, 1976, p. 16).

> It is fundamental to the metaphysical doctrine of the philosophy of organism, that the notion of an actual entity as the unchanging subject of change is completely abandoned. An actual entity is at once the subject experiencing and the superject of its experiences. (Whitehead, 1929/1978, p. 43)

Conceived as constituted by being in process, everything is imbued with purpose and feeling. This is another key aspect of a logic of relationship. Unlike the mechanism of Darwinian evolution, change and process are driven by and coemerge with meaning and value. "Emerging creatures, appropriating but surpassing the past in their becoming, seek aesthetic satisfaction and intensity of feeling" (Cobb & Schroeder, 1981, p. xi). Process philosophies evolving from Whitehead's original notions share this common panpsychist perspective (literally coming from the Greek "all-soul"). This perspective of relationship and meaning as fundamental being-in-process, infused with meaning and purpose as driving agents, is contrary to the classical Newtonian mechanical perspective that implies the basic units of being are inert, unconscious, atomic units. These perspectives also reflect that of the Romantics. Becoming, according to process philosophy, therefore, is a purposeful, self-creative process.

> Freedom becomes in Whitehead's metaphysics the principal ontological category defining all things which truly exist, all *res verae*. Indeed, Whitehead's category of creativity is the principal criterion by which what is actual may be distinguished from what is merely abstract and derivative. (Lucas, 1979, p. 15)

No event can be completely determined by its past although all events are also partially caused by previous experiences. As explained by Gragg (1976):

> Each moment exists in partial independence of all predecessors by virtue of an element of chance novelty and spontaneity in each occurrence; and this element of chance in each event means that it must be undeducible and unpredictable in determinate detail from all its predecessors combined. (p. 50)

The relationships among process, freedom, feeling, and creativity provide the basis for a logic of relationship and are consistent with quantum physics and evolutionary systemic change. Past experience, comprised of moments of becoming, are limiting or defining but do not determine organismic and social creativity and potential.

> An actual entity is at once the product of the efficient past, and is also, in Spinoza's phrase, *causa sui*. Every philosophy recognizes, in some form or other, this factor of self-causation in what it takes to be ultimate actual fact. (Whitehead, 1929/1978, p. 228)

An underlying logic of relationship emanating from process philosophy expands notions of learning and explores the curriculum as emergent and dynamic. Student learning, rather than controlled by methods of instruction and well-designed learning modules, is seen as choices and unfolding experiences, within a social context. Creativity and self-direction can work against the goals of education as well as toward them. In process education, creativity and self-direction are encouraged as lifelong learning and meaning-making are emphasized. The systemic nature of social relations similarly places a premium on all students having opportunities for creative self-direction and opportunities for working toward social as well as individual good. To be avoided, according to Whitehead, are "inert" ideas and disconnected facts.

The curriculum, in continual process, becomes an endless cycle of exploration and inquiry, alternating among romance, precision, and generalization, according to Whitehead. While Whitehead defines romance, precision, and generalization as stages of learning, he treats them more fluidly than linearly. Joy, creativity, excitement, enthusiasm, and interest drive learning in Whitehead's process approach to education and the rhythm of learning, like our own breathing, is in constant relationship among romance, precision, and generalization. Emphasizing process and evolving relationship, education, from a process perspective, is holistic, experiential, interest driven, creative, and meaningful. The becoming of educative experience inherits the whole of our past experiences and aims toward a meaningful and purposeful future. Whitehead's subtitle for *Aims of Education* was "The Seamless Coat of Learning." Connected and holistic, education cannot be separated from living.

Process philosophy marks a significant break with the modern intellectual tradition with its focus on the logic of domination and scientific rationality. By focusing on process and exploring *causa sui* or self-cause, by understanding the rhythm of education as fluid and dynamic rather than linear and cumulative, process philosophy is more consistent with emergence science and postmodernism.

Rather than substituting one Grand Narrative for another (replacing the dominance of analytic philosophy with the perspective of process philosophy), process philosophy acknowledges its own speculative nature. The speculative nature of process philosophy is important to avoid yet one more "panacea" to solve our educational woes!

Speculative nature of process philosophy

Whitehead (1929/1978) describes speculative philosophy as a process of inquiry. Traditional definitions of speculative philosophy highlight the integrative, contextual, and cross-disciplinary nature of inquiry characteristic of speculative philosophy. As described by Evans (1998):

> One dictionary definition of speculative philosophy is, "philosophy which constructs a synthesis of knowledge from many fields (the sciences, the arts, religion, ethics, social sciences) and theorizes (reflects) about such things as its significance to humankind, and about what it indicates about reality as a whole." (p. 58)

Whitehead (1929/1978) himself defines the agenda of speculative philosophy to interpret every aspect of human experience. The interpretation is recognized as historically grounded and perspectival rather than a search for hidden reality or ultimate truth. The method must be coherent and comprehensive, involving all facets of experience. Speculative philosophy, rather than fragmenting the philosophical endeavor into discrete categories of investigation, ethics, logic, epistemology, and ontology, recognizes the interconnection and holism of meaning-making as well as the evolution of ideas and theories and the interrelationships among theories, approaches to inquiry, and experience. Speculative philosophy is a systemic approach to inquiry, the search for patterns of meaning and relationship. Examining meaning structures and practices as they affect all aspects of human activity tests the coherence of speculative philosophy. Recognizing the shifting ground of theory and the relationship between theory and practice, speculative philosophy itself is in process.

Dewey's logic as inquiry is relevant as we further develop a logic of relationship. His theory of inquiry expands Whitehead's notions of experience and creative emergence and has implications for changing how we see learning and education.

The Circle's Middle

The ways of thinking entrenched in the modern mind-set prevent us from seeing solutions to our problems in new and innovative ways. Einstein is noted for saying "No problem can be solved from the same consciousness that created it. We must learn to see the world anew" (Wheatley, 1994, p. 5). The underlying logic of modernism has been one of domination and control. Solutions to problems from this logical point of view entail setting cause-and-effect chains in motion and employing management models that attempt to regain a sense of control. The metaphors of change that utilize management vocabulary point to their modernist origins in control—control over nature, control over people, control over ideas, control over change, and control over things. To "manage" a classroom is to "control" learning and behavior. To manage a company is to control the employees and profit margins. "Managing" school reform, therefore, has an underlying modernist assumption of control and predictability. To enact a management plan is to set a chain of events in motion that will result in desirable ends.

Dewey's logic as his theory of inquiry offers a different perspective of schooling, the curriculum and the relationships involved in the processes of public education. Together, Dewey's logic and Whitehead's process philosophy reveal a logic of relationship that avoids the dualisms of traditional bivalent "thing" logics, provide insights on moral action, and allow us to explore the dynamics of habit and innovation, learning and creativity, process and meaning.

Dewey used Kant's phrase "The Copernican Revolution," to describe his own philosophy as an approach to living and to the "secrets" of inquiry. Contrary to Cartesian doubt and a mind-centered philosophy, Dewey's approach to philosophy marks a significant break with the Western philosophical tradition and provides both a key for and a perspective on Frost's mystery of the secret as a metaphoric way of understanding the paradoxes of modernism.

> *We dance around in a ring and suppose;*
> *But the Secret sits in the middle and knows.*
> (Frost, in Lathem, 1969, p. 362)

At the end of *The Quest for Certainty* (1929), Dewey summarizes his approach couched within the historical tradition of intellectual thought.

> We have seen how the opposition between knowing and doing, theory and practice, has been abandoned in the actual enterprise of scientific inquiry, how knowing goes forward by means of doing.... The old center was mind knowing by means of an equipment of powers complete within itself, and merely exercised upon an antecedent external material equally complete in itself. *The new center is indefinite interactions taking place within a course of nature which is not complete, but which is capable of direction to new and different results through the mediation of intentional operations....* There is a moving whole of interacting parts; a center emerges wherever there is effort to change them in a particular direction. (pp. 290–291) [emphasis added]

The secret in the middle is an emphasis on relationship, process, change, and meaning. Expanding or transforming our circle of knowledge requires a process-oriented, relational logic. "We may gain the world by renouncing it, by passively losing self in the heart of what has neither form nor dimension" (Chardin, 1965, p. 21).

The theory of evolution was influential in Dewey's approach to and focus of philosophy. He rejected the Cartesian approach to philosophy and challenged the idea that philosophy should be about finding immutable truths. As Dewey described in *The Influence of Darwin on Philosophy* (1910):

> The conceptions that had reigned in the philosophy of nature and knowledge for two thousand years, the conceptions that had become the familiar furniture of the mind, rested on the assumption of the superiority of the fixed and final; they rested upon treating change and origin as signs of defect and unreality. In laying hands upon the sacred ark of absolute permanency, in treating the forms that had been regarded as types of fixity and perfection as originating and passing away, the "Origin of Species" introduced a mode of thinking that in the end was bound to transform the logic of knowledge, and hence the treatment of morals, politics, and religion. (pp. 1–2)

Dewey's own philosophy carried out the mode of thought begun by Darwin, emphasizing evolutionary relationship. He rejected the "fixed and final" and found the illusiveness of the secret to be assumed permanence. The "secret" evolves and emerges as an interaction in dialectical relationship and transformational process. Frost's conundrum of the secret in the middle can be understood from Dewey's perspective as a changing, evolving complex of interactions and relationships. "We sit around the circle and suppose" that we can find a set secret in the middle, "while the secret sits in the middle and knows" that the secret of the middle is precisely our evolving relationship with it.

The basic tenets of Dewey's logic include the interconnections among action, thought, interaction, and evolution. These interconnections are grounded in experience. Dewey's focus on experience, his logic of inquiry as pragmatic problem-solving, his aesthetic perception and moral action, and his ethics emphasizing care provide the continuum of colored lenses through which the kaleidoscope of his vision can be captured and the secret in the middle can be engaged.

Dewey's Logic as Inquiry

Guiding Dewey's process approach is his focus on experience as it relates to the evolutionary dynamics of interactions of living organisms with their environments. Experience, for Dewey, interactively shapes both an individual and the environment as the dialectic of situations is played out. For living creatures, problem-solving guides experience.

Dewey's logic as inquiry is driven by problem-solving and builds upon his central focus on experience. Thinking is shown to be more than a cognitive process but to involve creativity and intuition as instrumental in the problem-solving complex which drives experience. Meaning, for Dewey, is a fundamental aspect of the problem-solving interaction that drives growth and is derived from experience.

Thinking drives problem-solving even at the most basic level. Dewey's notion of thinking includes intuition, instinct, feelings, imagination, and creativity as well as cognition and integration. As described by Holder (1995):

> James and Dewey viewed thinking as a process emergent from and continuously controlled by noncognitive levels of experience, levels that include experiential structures such as emotion, habit, and imagination. According to them, thinking is one kind of highly structured experience that emerges from and retains in an essential and dynamic way less structured experience. Both James and Dewey rejected cognitivist models of mental operations, because such models mistakenly maintain that structure exists only in the so-called cognitive aspects of the mind. (pp. 11–12)

As we engage in problem-solving, passion and imagination as integral aspects of thought are involved as well as habit and logic as practical reasoning. Problem-solving, therefore, is a fundamental human activity, for Dewey, and thought, as a human approach to solving problems, includes multiple "ways of knowing" beyond the traditional perspective of logical thinking.

Moral action and aesthetic perception

By locating events as the basis for relationships among experience, thought, and culture, Dewey's logic as his theory of inquiry expands on Whitehead's process philosophy and his emphasis on relationship as the most basic reality. Dewey's aesthetics and value theory are important perspectives of his logic of inquiry and ultimate treatment of the individual, social, and environmental complexities grounded in his philosophy of process. Morality, valuing, and aesthetics are dimensions of action driven by meaning. Educating for moral action requires, according to Dewey, nurturing imagination. Imagination and reflective abstraction allow for the playing out of possibilities as we struggle for resolution in problematic situations.

Engaging the creative imagination and thoughtful deliberation of potentialities are often bypassed by thoughtless action, an overreliance on tradition, or a pattern of action defining our own habits of interaction. What is worse is that education often provides experiences that disempower students for enlisting their creative imaginations. As described by Fesmire (1995):

> There are times when we plod along in a horizon-less daze. At other times, we are ecstatically focused on a future end to the impoverishment of present imagination. In both cases, our dramatic rehearsals are abbreviated. When our

> imagination is contracted, we do not attend to the world's possibilities....[E]ach time we approach a project as an externally imposed burden to be borne, we scoop out just a little more dirt for a grave in which to bury our imaginations. (p. 53)

Dewey's aesthetics considers three ways the arts may engage us. The first is related to our specific interactions with objects of art. Experiencing and appreciating art, he argues, contribute meaning and value to our lives. Dewey also considers the instructional significance of art. Creative imagination is engaged in both the exploration and creation of objects of art. A third aspect of Dewey's approach to aesthetics is related to inquiry. While not distinct from the experiences and instructional significance of art, aesthetic perception, according to Dewey, pervades all perception and should be nurtured. It is only through aesthetic perception that moral acts are possible.

Dewey's notion of aesthetic perception involves the creative and reflective practice of examining potential ends-in-view to hypothesized action. Actualizing the potential ends-in-view is a creative act. Moral action is the selection of particular modes of action from those possibilities and is therefore intimately connected with creativity. The ends-in-view approach also recognizes the evolving yet indeterminate unending flow of relationship as the changes we implement affect our own environments. Long-term ends are not possible to enact other than as processes involving ends-in-view action. Moral action engages aesthetic perception to generate and select our interactive strategies and solutions to our problems. A sense of the interconnectedness of action and the multiplicity of relationships emanating from possible action is related to aesthetic perception and, when considered, also becomes a part of moral action.

Aesthetic perception also includes the feelings of closure, of having generated options and selected from among them. The notion of "flow" is associated with aesthetic experience as feelings of deliberation, unity, growth, balance, and awareness become associated with aesthetic perception. Aesthetic perception, therefore, is imbued with meaning, purpose, and value. Creative problem-solving involves the aesthetic experience. Aesthetics, therefore, is "the opening up of latent possibilities for growth, meaning, and fruitful action" (Fesmire, 1995, p. 56).

Aesthetic experience is tied to inquiry through moral action; however, not all aesthetic experiences implicate moral acts, as described by Fesmire (1995):

> *Any* decision that resolves a state of doubt has some aesthetic quality. This is why we are often taken in by our poor moral judgments. An intelligent moral decision deals sensitively with desires that demand attention rather than stifling them, thereby leading to inclusive and enduring rather than individualistic and fleeting consummation. The moral artist imagines ways in which others' lives can meaningfully develop alongside his or her own. (p. 56)

Moral action, therefore, is not rule-following. Following rules does not create a state of doubt and, therefore, does not require a decision. Without having to decide on the appropriate course of action while relying on the given rules, without having to consider the consequences of our actions, we are not engaged in moral decision-making. Moral action assumes the deliberation of possibilities and the consideration of others. As an aspect of aesthetic experience, moral action leads to an ethic of care or empathy that is particularly relevant to education.

Thus, moral action implies the nurturance of creative action and aesthetic experience. It also includes the assumption of the position of other and a deliberate selection of "ends-in-view" from the infinite possibilities of potential actions that considers the consequences and positionality of others. Developing a moral stance that values the impact of our actions on others, our environment, and ourselves exhibits an ethic of care. As such, moral education subsumes an ethic of care implicit in valuing.

The Heart of Relationship

Underlying Whitehead's process philosophy and Dewey's logic as inquiry is a logic of relationship with the following characteristics:

- Holistic being is constituted by experience
- Events or occasions of experience are primary or fundamental reality
- All being as occasions of experience are imbued with meaning, purpose, and freedom
- *Causa sui* or self-cause underlies all processes of change
- Inquiry guides action
- Morality, valuing, and aesthetics are dimensions of action driven by meaning
- Growth through practical inquiry and expansive relationships includes creative imagination, intuition, and feelings
- The systemic relationships of and interconnections among individuals, themselves complexes of events or occasions of experiences, can only be explored as patterns of relationships

The logic of relationship represents one perspective of recreating heart, of finding meaning and purpose to address the tensions and changes of the postmodern struggles we face. Dewey even addressed notions of the heart as interconnected, creative, and meaningful exploration and process. As described by Alexander (1995):

> The old use of "heart" is perhaps closer to [Dewey's] meaning than the modern idea of "mind." We must try to enter into the full meaning of situations, not merely a "cognitive" understanding of them. Our imaginative ability to shift from one perspective to another facilitates the process of communication and brings diverse perspectives into a *working* relationship, even though there may be a degree of *conceptual* vagueness as a result.... Instead of teaching people to develop merely their own powers of self-justification, we need to develop our capacities to hear others, genuinely to listen to them, ... [to develop] a capacity to take the other's point of view into our own and through it to enlarge our understanding of the situation. Only insofar as we have developed our ability to enter imaginatively into the worlds of meaning that human beings have constructed to make sense out of life, can we begin to listen to the richness of experience itself. (p. 88)

Dewey's logic as inquiry and Whitehead's process philosophy recreate heart by turning away from cognition as something that happens in minds. By naturalizing mind and describing how intelligence evolves, we are no longer isolated within our bodies worrying about evil geniuses who might be playing tricks on us. Dewey's theory of inquiry and Whitehead's process philosophy are important parts of our recreating heart and serve as a basis of a relational, emergent curriculum. Integral to a logic of relationship is an understanding of the dynamic qualities of our universe as a system in process.

Relationship, Morality, and Meaning

The logic of relationship is one part of the constellation of perspectives that comprise my postmodern logics. While the overlapping features of the logics of relationship, meaning, and systems may tempt us to consider them related or the same, these ways of seeing provide necessarily different yet complementary perspectives to guide action and offer meaning and moral purpose to our efforts to educate our children.

My own experiences and actions with my students and colleagues, first as a classroom teacher, then as a university professor, and now as an administrator, are guided by a logic of relationship. Respect and appreciation for other's needs for creative and purposeful action as well as an understanding of my own complexity of relationships and dictates for moral action have changed for me what seems most important about our shared educative experiences.

Karen Warren (1996) describes the experience of climbing a mountain. At first, she describes, she set out to conquer the mountain. As she struggled to climb the mountain, her belay rope many times kept her from falling as she found herself bruised and tattered by the mountain, dangling by her rope as she had time and again lost her footing. The next day, she chose to remove the safety rope and to go to the other side of the mountain, alone. She picked the hardest part but, rather than approaching the climbing of the mountain as a task, she changed how she thought of the mountain. She wondered how the mountain felt and decided to develop a relationship with it.

> I closed my eyes and began to feel the rock with my hands—the crack and crannies, the raised lichen and mosses, the almost imperceptible nubs that might provide a resting place for my fingers and toes when I began to climb. At that moment I was bathed in serenity.
>
> I began to talk to the rock in an almost inaudible, child-like way, as if the rock were my friend. I felt an overwhelming sense of gratitude for what it offered me—a chance to know myself and the rock differently, to appreciate unforeseen miracles like the tiny flowers growing in the even tinier cracks in the rock's surface, and to come to know a sense of *being in relationship* with the natural environment. It felt as if the rock and I were silent conversational partners in a long-standing friendship.
>
> I realized then that I had come to care about this cliff which was so different from me, so unmovable and invincible, independent and seemingly indifferent to my presence. I wanted to be with the rock as I climbed. Gone was the determination to conquer the rock, to forcefully impose my will on it; I wanted simply to work respectfully with the rock as I climbed. And as I climbed, that is what I felt. I felt myself *caring* for this rock and feeling thankful that climbing provided the opportunity for me to know it and myself in this new way. (pp. 26–27)

The power of the logic of relationship for me has been coming to know my students and myself in a different way. The subtle yet profound change that has occurred as I think of my students as relationships, not things, and consider the moral purpose of education has implications for education that may transcend the logic of domination.

Notes

1 My use of "he" and "his" is deliberate in this section to emphasize that science, the scientific methods, and the foundations of modernism were based on patriarchal thinking and concerns.

2 Whitehead uses the terms "rational" and "anti-rational" distinctly from the philosophical perspective of rationalism. He intends to imply the unquestioning acceptance of the scientific paradigm and modern approaches to understanding being makes the underlying perspective of modernism anti-rational or unquestioning.

CHAPTER FIVE
The Logic of Systems

In this chapter, I present the logic of systems as another aspect of our postmodern logics perspective. The logic of systems, originating in general systems theory, rejects the piecemeal approach of modernism and considers the form of systems per se. The commonalities of systems and the emergence of complex adaptive systems and chaos theories in the latter part of the 20th century challenge previous notions of logic and relationship in systems.

Traditional forms of logic cannot account for the complexity and relationship found in many systems. The sterility, abstraction, and formalism of logic, especially symbolic logic, is far removed from our everyday world of chance, accident, mistakes, misjudgments, creativity, and context. Nevertheless, I began my formal philosophical studies in the 1970s with the study of logic and was fascinated by the systematicity of it. My mathematics background facilitated how I approached and understood the meaning and significance of logic. Emerging theories of multivalued logics, fuzzy logic[1], and artificial intelligence attempted to expand the formalism of mathematics to human and machine thinking. Questions of intelligent action versus logical action were explored and the budding field of artificial intelligence captured my philosophical as well as science fiction interests.

Incongruities with the logical foundations of mathematics were apparent when attempts to explore the foundations of mathematics in the late 19th century focused on Frege's logic.[2] Russell's paradox described in a tangible way the implications of infinite sets for a mathematics based on extensions of Frege's set theory. The question Russell raised was:

"Does the set of all infinite sets contain itself?" This question is the 20th century equivalent to an entire class of paradoxes including the paradox of the Barber of Seville: "The barber of Seville shaves all and only those who do not shave themselves. Does he shave himself?" Surely logic as the foundations of mathematics should not lead to such problems!

Early 20th century constructivists and intuitionists approached the problems of the foundations of mathematics[3] by denying the most basic logical truth of mathematics: that either P or not-P must be true. While the logic of mathematics and the paradigms of argumentation, including scientific reasoning, rely on the Law of the Excluded Middle, it was certainly clear to me that for most experiences, context made the generalizable truth of "P or not-P" problematic. "It's raining or it's not raining" is not a statement whose truth is easily resolved. It may be raining in Tulsa and not raining in Wichita. The weather, schools, children, and most things of interest are complex (often parts of even larger systems) and seem intimately and inextricably related to context. Truth statements and valuation based on logical form no longer seemed relevant in a messy, interrelated, interconnected, relational world.

Russell, Whitehead, Wittgenstein, among others, wrestled with the problem of the foundations of mathematics until Gödel published his proof of the incompleteness of mathematics in the early 1930s. This proof recast all study of the foundations of mathematics and forever changed earlier ideas about mathematics as God's language of the universe. No mathematics, Gödel proved, could be complete without leading to contradictions. It now seemed apparent that mathematics could not capture our ideas about truth, which itself seemed to be a system of relationships. In contrast with many more naturally occurring systems, mathematics is fairly barren of the hope for generating new knowledge. Because mathematics is a closed system, meaning and truth are restricted by the given, small set of primitives and postulates. Newton's clockwork universe and the mathematization of our world are not rich enough to handle the complexity and evolving nature of the world around us and of which we are a part.

Multivalued logic offers the promise of better capturing the uncertainty, complexity, and interconnectedness of living, real-world experiences. It, too, is limited, however, by the mechanics of truth valuation. Standard and alternative logics based on truth-value claims lost their appeal to me, as did analytic philosophy in general, because of their inabilities to deal with complex, naturally occurring, living and evolving, complex systems. The problems and approaches of analytic philosophy and logic are incapable of addressing the complexities of the world of relationship, perspective, and meaning.

I left philosophy to be a mathematics and computer teacher. An epiphany for me was the rediscovery of process philosophy leading to an exploration of a logic of relationship. My next major breakthrough came with increasing awareness of the complexity sciences. Underlying most living, emerging, and complex relationships is a logic of systems that will be explored in this chapter. The Law of the Excluded Middle and the Law of Contradiction are rejected as foundational to a logic of systems, signifying a break with modernist logic and reasoning.

Science, Technology, and Tools

In their book, *The Knight's Move,* Loder and Neidhardt (1992) distinguish between science and technology:

> *Science* may be defined as a human activity that strives for *explanation* by close *observation* of its world of exploration within an interpretative framework that brings observation to the level of meaningful understanding.... *Technology* may be defined as the devising of actions taken for material advantage. Technology stresses the development of techniques by which one can manipulate the material order to further utilitarian ends. (p. 3)

Modernism conflates the roles of science and technology by confusing the tools of science for the meaning of science. The clock, as a fundamental tool of science in the modern era, became the predominate metaphor for the universe. By confusing the tool with meaning structures, however, science itself has become void of meaning and significance.

> A technological perception of science reduces the meaning of science to the manipulation of the material world. Thus, "science and technology" takes on an inflated mythological force, creating a truncated world view in which human control of the systematic relationship between ends and means suppresses issues of purpose and meaning. The only metaphysic that can come from a technological outlook is a closed-system determinism. (Loder & Neidhardt, 1992, p. 4)

Science as a meaning system becomes much more closely aligned with spirituality, relationship, and interdependence when the tools of science become relegated to method and science remains focused on meaning. When the tools become confused with their meaning, science becomes like mathematics; a closed, incomplete, barren endeavor. Formal logic, as a way to understand mathematics, relegates mathematics to manipulation of symbols when the tools of mathematics are the basis of that mathematics.

Maxwell and Einstein are examples of two scientists who have been accused of infusing their ideas about science with mystical ideas of transcendence, creativity, and spirituality. These accusations of mysticism come from the modernist perspective of science as the technical production of tools. From the perspective of science as meaning structures, however, the distinction between science and spirituality becomes blurred. Postmodern science, rejecting the tenets of the perfect predictability of the clockwork universe of Newton and recognizing the tools of science as useful meaning and interpretive frameworks rather than conveyors of truth, is open-ended, probabilistic, indeterminate, relational, and dynamic. As a system of meaning, the relationship between science and spirituality is not as clear as modernist notions of the objectivity of science suggest.

While the tools of postmodern science have changed from mechanical to electro-chemical, if we are not careful to separate science from its tools, the same reduction of meaning to closed-system thinking is a threat. This chapter explores how the complexity sciences have contributed to the meanings and metaphors of postmodern process theories about relationship, creativity, and emergence. The underlying logic of the complexity sciences is the logic of systems. We must be forewarned however, not to confuse the tools of postmodern science with the meaning structures of postmodern science. The logic of systems is a way of exploring and understanding our experiences and observations but must not be confused as representing some hidden truths of nature.

This chapter explores the science of complex adaptive systems and the meanings and metaphors inherent in postmodern science with their underlying logic of systems. The complaints of scientism and overt mysticism are quelled by distinguishing between science as meaning and the tools of science. The first sections of this chapter will describe the complexity sciences as meaning systems. The *Erwartungshorizonten* of the complexity sciences can be traced to systems theory, so the early sections of this chapter will provide the historical context of systems thinking and systems theory, and their relationship to systems logic. The middle section of this chapter will explore and develop complex adaptive systems as living systems. Understanding living systems from a systems logic perspective will extend our ideas about social systems, further explored in Chapter Seven. Chaos mathematics will be explored as a powerful tool for understanding and exploring complexity. It will also be argued that chaos techniques, as tools for understanding complex systems, must not be elevated to the point where we confuse the tools with the science. Chaos mathematics, however, offers powerful visual and practical metaphors for understanding complex systems. The Mandelbrot set, in particular, will be presented as a powerful way to understand schooling, learning, and classroom dynamics as complex adaptive systems in later chapters. The final section of this chapter will position complex adaptive systems in postmodern thought. The challenge of postmodernism as yet another "Grand Narrative" will be addressed in terms of systems logic underlying postmodern science.

General Systems Thinking

Systems theory, in general, shifts attention from physics and mathematics to biology. Questions of relationship, change, and process are explored from systems perspectives. In particular, the question "What does it mean to be living?" drives many of these approaches. The fragmentation of the sciences and emulation of physics as the paragon of scientific certainty are challenged as commonalities among the biological and social sciences are examined. Systems theory originated early in the 20th century, just as quantum physics, theories of relativity, and evolutionary biology were emerging.

Bergson (1911) was among the first to explore psychosocial relationships from a systems perspective, focusing on biological and naturalistic processes. He (as was Dewey) was born the same year that Darwin published *The Origin of Species* (1859). His writings were in conversation with and in opposition to the determinism of Darwin. The mind and intelligence, according to Bergson, are driven by creativity and intuition rather than intellect and knowledge per se. The goal of all life, according to Bergson, is creative action rather than knowledge or certainty. Intuition rather than analysis reveals the true reality or nature of the universe. According to his perspective, survival is related to creative action rather than genetic disposition.

The rejection of the mechanical and teleological perspectives of Darwinian evolution and Newtonian physics is characteristic of general systems thinkers who attempt to understand complexity, emergence, relationship, creativity, and change from post-Darwinian perspectives. Bergson's approach was primarily philosophic with a focus on psychosocial processes. Bertalanffy's approach, several decades later, emerged from the field of biology rather than psychology and challenged traditional scientific thinking from within the discipline of biology.

In the late 1920s, Bertalanffy[4] started to explore what he termed "organismic biology." General systems theory, more fully developed after World War II, evolved from generalizations of his organismic biological theory and utililized many of the process metaphors of evolution and ecology. Bertalanffy's interest in organismic biology was driven by his dissatisfaction with the mechanical metaphors of modern science. The holism and emergent properties of living organisms could not be understood from the fragmented and reductionistic perspectives of modern inquiry.

Bertalanffy also became interested in a general theory of systems that explained common features of a variety of different kinds of systems including living and nonliving systems, and biological and social systems. His general systems theory (GST) was formulated at the end of World War II. The goal was to consider the general patterns of relationship and organizational features of a variety of physical, chemical, living, and social systems. He was concerned that the fragmentation of the sciences was preventing exploration of some very fundamental principles. Thus, GST became as much a method as a philosophy of inquiry and transcended several domains of inquiry.

> Physics is still the paragon of science, the basis of our idea of society and our image of man. In the meanwhile, however, new sciences have arisen—the life, behavioral, and social sciences. They demand their place in a modern world view.... Now we are looking for another basic outlook on the world—*the world as organization.* Such a conception—if it can be substantiated—would indeed change the basic categories upon which scientific thought rests, and profoundly influence practical attitudes. (Bertalanffy, 1968, p. 187)

A decade later, Bertalanffy joined forces with Boulding (1956) (economist), Gerard (neurophysiologist), Miller (psychologist), Rappoport (mathematician), and Margaret Mead (anthropologist) to form the Society for General Systems Research in 1954. The goals of the Society for General Systems Research, now known as the International Society for the Systems Sciences (ISSS), were to

> (1) investigate the isomorphy of concepts, laws, and models in various fields, and to help in useful transfers from one field to another; (2) encourage the development of adequate theoretical models in the fields which lack them; (3) minimize the duplication of theoretical effort in different fields; (4) promote the unity of science through improving communication among specialists. (Bertalanffy, 1975, p. 155)

As described on the Internet home page of the ISSS:

> The founders of ISSS felt strongly that the systemic (holistic) aspect of reality was being overlooked or downgraded by the conventional disciplines, which emphasize specialization and reductionistic approaches to science. The founders stressed the need for more general principles and theories, and sought to create a professional organization that would transcend the tendency toward fragmentation in the scientific enterprise.
> (http:// www.isss.org/draft.htm)

Banathy (1998) distinguishes between systems theory, systems philosophy, and systems methodology. Systems philosophy, according to Banathy:

> [A]sks the question, "How can we understand systems?" With the perspectives of systems philosophy, we look at the world in terms of facts and events in the context of wholes, and we understand them as integrated sets purposefully arranged in systemic relations. In contrast to the analytic, reductionist, linear, single cause-and-effect view of the philosophy of classical science, systems philosophy brings forth a reorganization of ways of thinking and knowing perceived reality, a view manifested in synthetic, expansionist, dynamic, and multiple/mutual causality modes of thinking and inquiring, how things work more than what things are. (p. 6 of 16)

He delineates the conceptual foundations of GST by comparing systems orientations with classical, traditional approach to science. (See Table 5.1) The focus of systems inquiry is on multiple/dynamic interactions, patterns of relationship, and the whole.

The logic of systems includes the philosophy of systems, theory of systems, and method of systems to comprise a worldview, an investigative orientation, and an emerging methodology. The goals of inquiry for understanding and activity distinguish the logic of systems from the logic of modernism and implicate or acknowledge the interactive nature of knowing. From a systems perspective, there is no longer the possibility of an objective observer or an Archimedian stance.

The fragmentation of the sciences and increased specialization within the sciences is a function of the underlying beliefs of classical physics. Systems approaches attempt to understand, within and across disciplines, structure, organization, and process. Emergence, increased complexity, and system viability are aspects of systems that are explored from these various perspectives. The goal is not better knowledge of individual systems but understanding of how systems qua systems operate. Exploring commonalities among biological, social, economic, chemical, and neurological systems, in particular, drove early systems research. Angyal early on explored the logic of systems as distinct from the logic of classical scientific reasoning and process.

	Perspectives of Classical Science	Perspectives of Systems Science
Focus	Single variables, parts, linear, relationships	Multiple/dynamic interactions, patterns of relationship, the whole
Modes of Inquiry	Analysis, reductionism, entity	Synthesis, expansionism, emergence, process
Reasoning	Cause-effect, determinism	nondeterministic, purpose, meaning
Rule	Objectivity, observer detachment	Observer involvement and influence
Goal	Prediction, identity, knowledge	Understanding, activity
Control	Goal-driven, negative feedback, adjust for error	Goal-driven (adjust for error), positive change of goals

Table 5.1: Comparison of Classical and Systems Sciences. See Banathy (1998)

Underlying Logic

Angyal (1941) argues that the logic of systems is different from traditional scientific reasoning and the logic of causality. He delineates between systems logic and traditional logic by noting the following differences:

1. A relation (in traditional logic) requires two and only two members (relata) between which the relation is established. A complex relation can be always analysed into pairs of relata, while the system cannot be thus analysed.... [T]he type of connexions (sic.) in a whole is very different from the connexions which exist in an aggregate. The term "system" is used here to denote *a holistic system*. Further, in using this term we abstract *constituents* ("elements") and refer only to the *organization* of the whole. Thus "system" for our discussion is holistic organization.
2. ...A relation (in traditional logic) requires an aspect out of which the relationship is formed. Two objects can be related to each other, for instance, with regard to their color, size, or weight.... The members of a system, on the contrary, do not become constituents of the system by means of their immanent qualities, but by means of their distribution or arrangement within the system.
3. Although the dimensional domain is a necessary condition for both relationships and systems, the function of the dimensional domain is different for these

> two logical genera. The role of the dimensional domain for a relationship (in traditional logic) is merely disjunction of the relata. But the role of the dimensional domain ends here. The domain itself does not enter into the relationship. For instance, if two colors are separated in space a comparison between them can be made without any further reference to space. The dimensional domain is more intimately involved in the formation of systems (in systems logic). Here the dimensional domain not only separates the parts, but it participates in the formation of the system. *The system is dimensional. A system is a distribution of the members in a dimensional domain.*
> 4. ... Dealing with relations and dealing with systems involve two different logical manipulations to which two psychologically quite different processes may correspond. In the recent past there has been much rather inconclusive discussion concerning the possibility of two different processes of knowing: explanation and understanding.... The difference between the two concepts ... is probably that explanation refers to relational thinking, understanding to system thinking. Relational thinking aims at the establishment of the direct connexion (sic.) between two objects.... In systems thinking the task is not to find direct relations between members but to find the superordinate system in which they are connected. (pp. 18–24)

Traditional logic lends itself to Newtonian examination. From this mechanical perspective, the number of pairs of relationships determines the complexity of the system's logical space. Systems logic, on the other hand, explores the complexity of a system as the interconnectedness and patterns of relationship within the system. The system can only be understood from the perspective of interdependence, organizational structure, and evolving dynamics.

A logic of relationship alone, however, is not sufficient for understanding the complexity of relationships and the organizational dynamics of systems as a whole. The focus on relationships, not the parts themselves, must also include the organizational dynamics. "*In aggregates it is significant that the parts are added; in a system it is significant that the parts are arranged*" (p. 26). The aggregate as the sum of its parts is an entirely different process, according to Angyal. A system does not obtain its identity and characteristics from the aggregate of its parts or the relationship among pairs of parts of the whole. There is a unity or *unitas multiplex* that is defined by systemic organization that is greater than the composite of the individual parts or the pairwise comparison of parts in relationship. The logical formulation of the system itself is greater than the particular or individual organizational relations among the parts. Systems theory is about the exploration and examination of the organizational properties of systems that lend coherence and meaning rather than the particular arrangements found in individual instances of systems. This shift represents an underlying systems logic.

Of particular interest are dynamic or open systems. At the time in which Angyal was writing, understanding system organization in dynamic living systems was a new science. Bertalanffy made the first distinction between open and closed systems. Prigogine made tremendous progress in understanding the dynamics of open systems in the area of thermodynamics twenty years later while Maturana and Varela utilized many of these ideas in formulating their ideas about living systems as self-producing, autonomous, complex unities starting in the late 1960s. These distinctions and additions to the discussion on complexity and unity within dynamic, open, living systems helped evolve GST to complex adaptive systems theories. These distinctions and transitions will be described below.

Systems Theory and New Science

Systems thinking is used in a variety of scientific and nonscientific realms. The use of the phrase "systems theory" in no way should imply there is a single, unified theory about how systems operate. The logic of systems, as described above, includes the methods, philosophy, and theory of systems. The physicist Fritjof Capra (1996) characterizes systems thinking as a move away from the mechanistic and reductionistic thinking of modern science to the holistic or organismic thinking of much of 20th-century science.

> The basic tension is one between the parts and the whole. The emphasis on the parts has been called mechanistic, reductionistic, or atomistic; the emphasis on the whole holistic, organismic, or ecological. In twentieth-century science the holistic perspective has become known as "systemic" and the way of thinking it implies as "systems thinking." (Capra, 1996, p. 17)

Although initiated by biologists, systems thinking today is found in most branches of the empirical sciences. Rejection of mechanistic models encompasses more than the scientific perspective, however. Romanticism in art, literature, and philosophy during the late 18th and 19th centuries similarly marked a shift in perspective to holism and mutual interdependencies, as discussed in Chapter One. Postmodernism, in its variety of forms, is a challenge to the fragmentation and isolation of modern thinking. Thus, constructive, deconstructive, existential, and neo-Marxist postmodern perspectives all share a common concern with the fragmentation, mechanism, and isolation associated with modernism.

Self-organization as dissipative structure is a core aspect of the evolutionary dynamics of open systems and complex adaptive systems theory as it has emerged in the late 20th century. An essential aspect of system structure from this perspective is holistic relationship. Thus, as Capra (1996) summarizes:

> A system has come to mean an integrated whole whose essential properties arise from the relationships between its parts, and "systems thinking" the understanding of phenomenon within the context of a larger whole. This is, in fact, the root meaning of the word "system," which derives from the Greek *synhistanai* ("to place together"). To understand things systematically literally means to put them into a context, to establish the nature of their relationships. (p. 27)

While the systems perspective provided new insights in the field of evolutionary biology, the paradoxes of quantum physics were reinterpreted from a systems perspective to understand the behavior of quanta and the importance of systems relationships in observation and being. Beyond our own interpretive influence of observing quantum interactions, quanta themselves, from a systems perspective, have meaning only in connection and relationship with each other and in comparison with the system as a whole. Gestalt psychology, ecological studies, quantum physics, and organismic biology were the early fields of influence on the development of complex adaptive systems thinking.

At risk of reducing systems thinking to a list of criteria, there are several characteristics of systems approaches that can be identified and summarized from the above discussion. First, systems thinking focuses on relationship and structure rather than individual parts or the sum of pairs of relationships. Relevant properties of the system cannot be discovered by examining isolated aspects of it; the systemic nature is destroyed when investigated piece-meal. The important features of a system are precisely the organization and relationship of its various parts in relation to the whole.

A second consideration of most systems approaches is levels of organization. During the analysis of organization and relational structure, perspectives and hierarchies of organization are observed. Levels of organization within structures are seen in most complex systems; understanding the process of organization at different levels and between levels includes exploration of the dissipative, emergent, generative properties of systems.

A third important component of systems thinking is context. In addition to the primacy of relations within the system, system dynamics are intricately connected with both historical dynamics and external environments. Systems thinking requires exploration of system relationships from the context of its own history and environmental relationships.

New science examines systems as self-generating, historical, contextual, relational, and interdependent. The emphasis on system relationship, the focus on transformative process, and the implicit notions of self-generation and creativity of the self-actualizing process are key components of the logic of systems underlying New Science. New Science, as described, may be a revolutionary way to approach and extend our understandings and how we interact with each other and in the world. As suggested by Rennie (1999): "real revolutions, however, take place inside our brains, when science helps us to understand the world anew" (Rennie, 1999, p. 1).

The logic of systems displaces the binary focus of scientific inquiry by recursively exploring system interactions and tracing system histories within a natural context. Complex adaptive systems approaches to science reveal the methods, philosophy, and theory of systems logic.

Complex Adaptive Systems

Several disciplines contribute to complex adaptive systems research. Originating as studies of complex systems, complex adaptive systems research is a late 20th-century approach to understanding living systems. Complex adaptive systems research emerged primarily from studies in nonequilibrium physics and dynamical systems.

Prigogine studied the irreversibility of many chemical reactions in thermodynamics, a branch of nonequilibrium physics, while Lorenz, among others, initiated the field of dynamical studies of the weather. Nonequilibrium physics studies systems at *far-from-equilibrium* states while dynamical systems research explores the amplification of small changes leading to dramatic effects, known as *nonlinearity*, in many types of systems. Complex adaptive systems research applies many of the findings of complex systems studies to living systems, exploring how far-from-equilibrium states and nonlinearity may affect living systems. Self-creativity or autopoiesis seems to be the response of living systems to these states.

The next section will develop complex adaptive systems from its origins in thermodynamics and dynamical systems research. Complex adaptive systems research, as an approach to understanding living systems, offers a perspective, based in systems logic, for exploring teaching, learning, and the complex array of relationships associated with schooling.

Nonequilibrium physics and far-from-equilibrium states

Prigogine was awarded the Noble Prize in chemistry in 1977 for his experiments and theories in thermodynamics resulting in his theory of irreversibility. Reactions to system perturbations at far-from-equilibrium states of certain chemical systems seemed to result, not in system breakdown, but in *system reorganization and increased complexity*.

The irreversibility phenomena seemed inconsistent with standard views of deterministic change and knowledge. Laplace's application of Newtonian reasoning had suggested that, if we had better knowledge of initial conditions, we could predict with greater certainty both future and past events. Irreversibility abolishes such possibilities. Similarly, increased complexity challenges Newton's notions of entropy or the idea that as systems "wear down" they lose energy and become more homogeneous rather than more complex.

Prigogine's experiments also explored the possibility of the emergence of life from chemical processes. "Can chemical processes provide us with the key to the difference between the behavior of a crystal and that of a cell?" he asked (Prigogine & Stengers, 1984, pp. 131–132). An answer to this question leads to the distinction between the classical theory of chemical kinetics and emergence theories used to describe properties inherent in thermodynamic reactions.

> The classical theory of chemical kinetics is based on the assumption that the rate of a chemical reaction is proportional to the concentrations of the reactants taking part in it (the sum of the parts) (p. 132).... If left to itself, a system in which chemical reactions occur tends toward a state of chemical equilibrium (p. 133).... In contrast with close-to-equilibrium situations, the behavior of a far-from-equilibrium system becomes highly specific. There is no longer any universally valid law from which the overall behavior of the system can be deduced. Each system is a separate case; each set of chemical reactions must be investigated and will produce a qualitatively different behavior. (pp. 144–145)

In such systems, according to Prigogine, self-feedback constructively contributes to reorganization. These "catalytic loops" affect the system in a nonlinear, nonpredictable way. The process of system feedback at far-from-equilibrium states may be the nonreductionistic answer to the question of evolution. As described by Wheatley (1994):

> [T]he dissipative structures that Prigogine studied demonstrated the capacity of living systems to respond to disorder (non-equilibrium) with renewed life.

> Disorder can play a critical role in giving birth to new, higher forms of order. (p. 11)

Prigigone's discoveries that far-from-equilibrium states of open systems often result in an exchange of matter and energy resulting in feedback to the system that may create nonlinear, irreversible, and systemically unique effects ties his studies in thermodynamics with dynamical studies of other open systems.

Dynamical systems and nonlinearity

Lorenz's study of the weather similarly explored the complexity of systems at far-from-equilibrium states resulting in amplification of small changes to the system. Lorenz's early attempts to model the weather using nonlinear differential equations showed promise in the early 1960s of capturing and advancing our understanding of the dynamics of weather. These efforts were guided by the hope of some day being able to predict and ultimately control the weather.

Using an analog computer, Lorenz set his model in motion. By allowing his model to run over long periods of time, Lorenz could represent months of weather data. The now prophetic story is that, on one particular day, he was interested in a particular portion of the weather modeled by his computer, so he reset the variables to values representing those initial conditions and reran his model from that point. He went for a cup of coffee, returning an hour later to find his model was completely different from the previous model of the weather generated with those same values. Surely there must be a computer error! He checked the connections and the equations. What he found was that the printout he had used to reenter the initial values rounded the values off to three decimal places while the computer used six in its calculations. Realizing that he had introduced a small change in the initial conditions of his model starting at that intermediate point, he concluded that weather was a sufficiently complex system that even these relatively minute changes in initial conditions produced dramatic affects in subsequent system dynamics.

Sensitivity to initial conditions and *nonlinearity* of effects of small changes challenge the notions of predictability and order of systems. This "butterfly effect" suggests that small changes in complex systems may affect dramatic and unpredictable changes in the system, just as a butterfly flapping its wings in Hong Kong may in some way influence a change in the weather in New York City days later. Because complex systems are interdependent webs of relationship and connectivity and cannot be studied piecemeal, a different approach to understanding the dynamics of complex systems was needed. Early pioneers in chaos dynamics tried to understand and represent the dynamics of complex systems.

Chaos theory and complex system dynamics

James York provided the first mathematical definition of chaos in 1975. Trying to understand the behavior of complex systems like the weather, he described the behavior of Lorenz's equations as chaotic. Using representations of the behavior of systems called "phase space diagrams," York determined that if systems transitioned through different states, they eventually would become "chaotic." These states, known as phase doubling, showed how complex system dynamics, at critical points, changed from steady states to patterned repetition to unpredictable, chaotic, yet patterned behavior. The behavior of systems was represented using computers and the same techniques Poincaré, a 19th century mathematician, had used to understand the effects of gravity on three interacting bodies. Some point to Poincaré's early attempts to solve the three-body problem of celestial mechanics as the roots of chaos theory (Ekeland, 1990).

The three-body problem results from the increased complexity that occurs when more than two celestial objects interact in a systemic way. Keplar's three laws of motion were developed in the 16th century to describe the motion of the planets. Newton systematized the motion of the planets by postulating the existence of a "hidden force," namely gravity. Newtonian mechanics was used for several centuries to predict the interaction of bodies in space as well as on earth. So successful was his mathematization of these relationships, and so ingrained was the assumption of the lawlike nature of the universe, that our modernist assumptions, as expressed by Laplace in the 19th century, are that, if the initial conditions of any (mechanical) system are known in sufficient detail, all prior and future relationships can be determined.

Poincaré, at the end of the 19th century, took issue with the determinism of Newtonian mechanics and the quantitative model upon which it was based. He developed qualitative approaches to understanding the dynamics of systems and recognized that sensitivity to initial conditions and nonlinear dynamics made long-range prediction impossible. In fact, in addition to the inherent variability and indeterminism of complex systems, sensitivity to initial conditions and nonlinearity also affect our computational potential.

While exploring the three–body problem, Poincaré challenged the assumption made by Newton that what seemed to work for two bodies would work for three. This challenge brought up the notion that the complexity of the system, with the addition of one more interacting force, becomes so great that we can no longer use Newton's equations to understand the behavior of these systems. Thus, while Newtonian mechanics seemed to work for two bodies, when three bodies were involved, the behavior of the system could not be determined.

The techniques Poincaré developed for examining the dynamics of bodies in motion focused on the relational characteristics of elements in the system and the behavior of the system as a whole. His two-dimensional mappings of the periodic trajectories of planets in orbit anticipated many of the techniques used by York, Henon, and others almost a century later when computers could be utilized.

Despite the lack of computing and modeling power, the complexity of the interactions of dynamical systems was apparent to Poincaré. Using a technique that represented the behavior of systems as recursive over time, Poincaré attempted very different models of system behavior. The recursive representations developed by Poincaré later became refined as phase space diagramming. Phase space diagramming, we will see, treats time and system behavior very differently than traditional Cartesian plots of system behavior. Cartesian plots (as the way we learned to represent functions in elementary algebra) could not capture the complexity and behavior of dynamical systems. Using early phase-space diagramming techniques, it became evident to Poincaré that in some situations, where it seemed a system was behaving very erratically (such as three planets interacting), the behavior of the system actually exhibited a hidden order and regular patterns of behavior. These systems were stable albeit unpredictable. Other systems became chaotic. When these relational mapping tools were utilized, patterns of organization were observed that were hidden when the traditional Cartesian mapping approaches were used.

Henon, in the 1960s, was among the first to explore the behavior of systems using a computer to model the dynamics of the systems he was studying using a similar relational approach. He showed, however, that, although there was regularity and therefore limited predictability in some systems, there was no possibility of repeatability and therefore no hope for complete predictability. There were, however, areas of "strange attraction," where the system seemed to behave in a regular if not completely predictable way. These findings corroborated the discoveries made by Lorenz around this same time. Sensitivity to initial conditions, constrained order, patterns of relationship, and "strange attractors" seemed to be common of many dynamical systems in many different fields.

The mathematics used to explore behavior of complex systems is recursive, relational, nonlinear, and dynamic rather than algorithmic and static and is used to model rather than predict system behavior. By using recursive procedures, where the state of the system is fed back into the system to understand the next state, these mathematical techniques more adequately capture the emergent and relational properties of systems over time.

The examples below show how these relational techniques have implications for our notions of time that are more consistent with theories of relativity and deny that time is an absolute. They also show how the dynamics of this simple system can be understood very differently when relational approaches are used.

When we consider the behavior of a simple pendulum, we recognize that, once set in motion, the pendulum eventually comes to a resting place, called equilibrium, and that with each consecutive cycle of the motion of the pendulum, the bob never quite returns to its maximum swing. Each swing becomes shorter until equilibrium is achieved.

A Newtonian representation of the behavior of the pendulum treats time as a continuous, independent dimension. Using a time-plot, time is represented as the independent variable on the x-axis and position plotted along the y-axis as the dependent variable. This kind of representation is the one most of us learned in high school mathematics.

In the Newtonian representation of the behavior of the pendulum below (See Figure 5.1a), the weight is held at point A, time zero, at its maximum distance from the equilibrium state. As the bob is released, it swings smoothly, progressing downward to the equilibrium position B, back up the other side to about the same distance as it was released, represented by point C on the graph, then back down to the equilibrium position at D, back up, but not quite as far as where it was released, at E, and so on, until it finally comes to rest at equilibrium, represented by point F.

Using the techniques first described by Poincaré and developed by Heron, Smale, and others in the 1960s, a phase-space diagram treats time not as an independent variable but as a dynamic variable. Each small change in the movement of the pendulum is fed back into the system so motion itself is a system rather than a function of time as an independent variable. (This approach eliminates Zeno's paradoxes of motion and addresses Hume's concerns of change over time by treating changing states in relationship to previous states rather than treating change over time.)

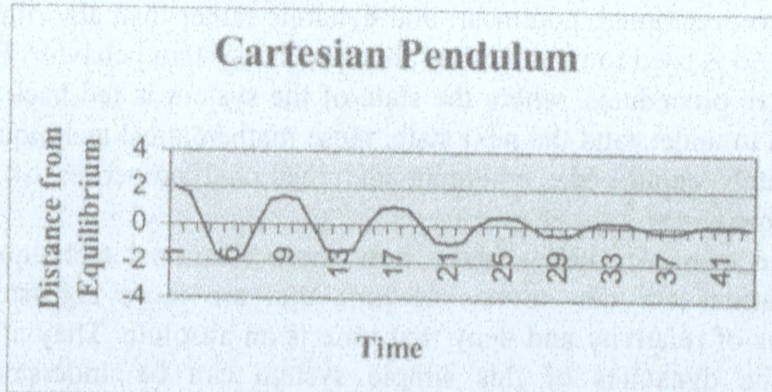

Figure 5.1 a Cartesian representation of the motion of a pendulum

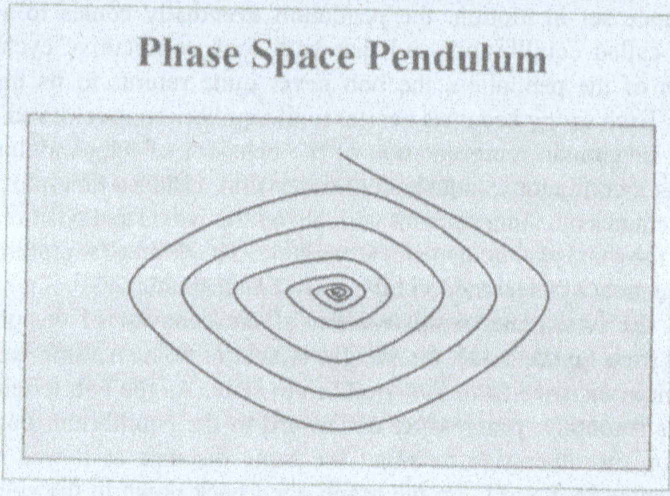

Figure 5.1 b Phase space representation of the motion of a pendulum

The phase-space diagram (see Figure 5.1b above) shows the behavior of the system as it progresses from the initial release of the bob to the final equilibrium state. The dimensions of the diagram are position or distance from the equilibrium point and speed or velocity of the bob. Tracing the diagram also gives an understanding of the direction of motion. Therefore, at point A, the bob is the furthest from the equilibrium point. As the bob is released, it travels downward at maximum speed, through the equilibrium point B, and back up to about or the same distance from the equilibrium point, C. At C, the direction of the bob changes, traveling back down toward the equilibrium point, so at point C, the bob is momentarily at rest and therefore has zero velocity. As the bob travels back through equilibrium, the direction, in reverse from the transition from A to B, is represented on the negative access, reaching a maximum velocity at point D. As the bob continues back up, it doesn't quite return to its position of release, represented by point E. The bob swings back and forth until it finally comes to rest at point F.

This simple change in representation of the behavior of the pendulum may not appear particularly advantageous. However, in open systems, where entropy does not eventually wear the system down, using phase-space diagrams can produce a vision of the dynamics of complex systems only imagined by Poincaré previously. Lorenz's butterfly effect (see Figure 5.2 below), took on new life when the phase-space diagram revealed the dynamics of the system seemed attracted about two stable yet unpredictable states. These "strange attractors" clearly revealed and reflected our intuitions about the weather. We recognize certain patterns in the behavior of the weather; however, for any particular day, we cannot predict the kind of weather we will have. The phase-space diagrams seem to capture the sensitivity to initial conditions, nonlinearity, emergence, and complexity while also revealing hidden order and constraint within the seemingly random behavior of the system.

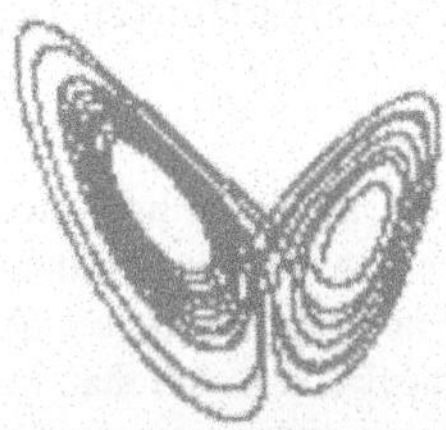

Figure 5.2 Lorenz butterfly

As these techniques were being developed for modeling dynamics of open systems, a geometry of recursion was being developed by Mandelbrot. The geometry of relationship not only formalizes the techniques of chaos, but also offers visual, artistic, creative, and analytic perspectives of complexity and the emerging order often found in open systems.

Chaos mathematics

Although trained as a mathematician, Mandelbrot rejected and was not particularly successful at formal mathematics while studying in

France as a Polish refugee. He was interested in a mathematics of relationship, not a closed-system mathematics, and approached his studies of mathematics intuitively and geometrically. Because of his unorthodox approaches and unwillingness to conform to traditional analytic approaches, and because the areas he was interested in studying were real-world, practical applications, Mandelbrot is described by Gleick (1987) as an "outsider, taking an unorthodox approach to an unfashionable corner of mathematics, exploring disciplines in which he was rarely welcomed, hiding his grandest ideas in efforts to get his papers published, surviving mainly on the confidence of his employers" (pp. 86–87). While the mathematics of the 20th century were largely constrained by axiomatization and formalism, Mandelbrot rebelled and insisted on exploring "real problems" from intuitive, geometric, recursive perspectives. After leaving France, he ended up in the United States working in the pure research division of IBM.

One problem that intrigued Mandelbrot was the pattern of interference in telephone communications. This was a significant problem for IBM since electronic communications were transmitted through telephone lines and interference affected transmissions. Mandelbrot examined transmissions for different lengths of time: for 60, 30, 20 seconds, and so on. Common sense would suggest at some short enough interval that the noise would not be found or would be continuous. This, however, was not the case. At all different time intervals, a common pattern of noise was observed.

> Within any burst of errors, no matter how short, there would always be periods of completely error-free transmission. Furthermore, he discovered a consistent geometric relationship between the bursts of errors and the spaces of clean transmissions.... In particular, it meant that, instead of trying to increase signal strength to drown out more and more noise, engineers should settle for a modest signal, accept the inevitability of errors and use a strategy of redundancy to catch and correct them. Mandelbrot also changed the way IBM's engineers thought about the cause of noise.... Mandelbrot's scaling patterns suggested that noise would never be explained on the basis of specific local events. (Gleick, 1987, pp. 91–92)

Mandelbrot's geometry of relationship seemed to capture the indeterminacy of mathematics itself while discovering something very significant about signal-to-noise ratios in electronic transmissions. No matter how much the transmission is amplified, there is a similarity, across scales, of noise to clear transmission and that patterns seem to occur across these scales that uniquely represent the dynamics of the system.

Mandelbrot tackled other diverse problems such as the flooding of the Nile and the fluctuation of cotton prices. What he discovered in these other studies, as with the noise patterns he observed in electronic transmissions, was that patterns seemed to emerge across scales revealing the system dynamics; systemic behavior seemed to repeat no matter how fine the gradation of data; and that system dynamics can be captured through the geometry of recursive presentation. Rather than trying to eliminate the "noise," control the system, or predict system behavior, Mandelbrot developed a geometry to explore the patterns that emerged. Chaos geometry therefore was invented by Mandelbrot to explore the irregularities in "real" problems, in the "real" world, rather than in the idealized world of perfect Platonic forms or abstract mathematics.

> The new geometry mirrors a universe that is rough, not rounded, scabrous, not smooth. It is a geometry of the pitted, pocked, and broken up, the twisted, tangled, and intertwined.... The pits and tangles are more than blemishes distorting the classic shapes of Euclidean geometry. They are often the keys to the essence of a thing. (Gleick, 1987, p. 94)

In developing the mathematics of recursion to explore system relationships, Mandelbrot reconceptualized the notion of dimensionality to try to capture the relational aspects of patterns across scales according to self-similarity characteristics. The structures were named "fractals" in 1975.

> Mandelbrot decided he needed a name for his shapes, his dimensions, and his geometry. His son was home from school, and Mandelbrot found himself thumbing through the boy's Latin dictionary. He came across the adjective *fractus*, from the verb *frangere*, to break. The resonance of the main English cognates—*fracture* and *fraction*—seemed appropriate. Mandelbrot created the word ... *fractal*. (Gleick, 1987, p. 98)

The fractal structure seemed, in many cases, to provide useful information, especially for structures or systems that exhibited irregularities, breaks, and branching. Quantitative measures such as size or distance were reinterpreted by Mandelbrot's chaos mathematics to include qualitative, relational features and to understand or model system behavior and characteristics. The structure of many biological systems like the lungs or the cardiovascular system can be more meaningfully discussed using fractal geometry. The fractional dimension and branching structure of the cardiovascular system, for example, explains how the expansive, complex network of blood vessels can extend over most of the body without taking up much space in the body.

> As a matter of physiological necessity, blood vessels must perform a bit of dimensional magic. Just as the Koch curve, for example, squeezes a line of infinite length into a small area, the circulatory system must squeeze a huge surface area into a limited volume. In terms of the body's resources, blood is expensive and space is at a premium.... [I]n most tissue, no cell is ever more than three or four cells away from a blood vessel. Yet the vessels and blood take up little space, no more than five percent of the body. (Gleick, 1987, p. 108)

Underlying the complexity of fractal structures is often simplicity and order. Simple dynamics can lead to complexity, similarity across scales, and a dimensional space that seems to defy traditional notions of one-, two-, and three-dimensional objects. Fractal geometry, therefore, provides a way for understanding complex systems where hidden order, similarity across scales, and change often emerge. "In the mind's eye, a fractal is a way of seeing infinity" (Gleick, 1987, p. 98) and "To see the world in a grain of sand" (Blake) capture the imagination, inspiration, and perspective of fractal geometry.

The complexity sciences and chaos mathematics can contribute to our discourse about relationships and challenge the foundations of modernism build upon the mathematization of nature. Embracing complexity, understanding relationship, and accepting probabilistic relationships and strange attractors rather than simple causality or certain knowledge, can move us beyond the paradigm of the clockwork universe inherent in the modern perspective. Mathematical discourse can be expanded to broaden our visions of relationship, combat notions of mechanistic determinism, reconnect us and our students with a mathematics of pattern and relationship, explore complex and dynamical systems, and embrace intuition and experience as meaningful in mathematical investigations and explorations. Explored from a systems logic perspective, the mathematics of chaos, nonequilibrium physics, and dynamical systems analyses can transform our old ways of looking at living systems, including children and social organizations. Complex adaptive systems theory explores an understanding of living systems from these various perspectives.

Living systems

Maturana (1980) explored the question: What does it mean to be living? As a biologist, this is the most fundamental question he could ask and one that he felt compelled to answer for his students. Approaching the problem of life is both a mandate of biology and a requisite of psychology. Before we can understand learning, we must understand the relationship among intelligence, creativity, and action, and a corequisite to these understandings is an understanding of life.

There traditionally have been three approaches to this question. The first approach, *vitalism*, was the perspective advanced by Bergson (1911) at the beginning of the 20th century. *Élan vital* was postulated as the force that gives any biological system life. Thus, the presence of this force characterizes the system as living and living things are distinguishable from nonliving entities precisely because they have the vital quality of life. The "vital force" of life, however, cannot be explained without involving a supreme force, being, or god who could bestow life on the inanimate. More difficult to accept, however, was the circularity of defining living things as those things with the quality of life.

A second approach to the question "What is life?" builds upon systems ideas. As described by Mingers (1995):

> [S]ystems theory, with the development of concepts such as feedback, homeostasis, and open systems, paved the way for explanations of the complex, goal-seeking behavior of organisms in purely mechanistic terms.... While this was a significant advance, such mechanisms could equally well be built into simple machines that would never qualify as living organisms. (p. 9)

Thus, if reduced to mechanistic relations, systems theory alone does not seem sufficient for explaining the emergence of life. The dissipative structures studied by Prigogine were chemical reactions at far-from-equilibrium states. While Prigogine hypothesized how life might emerge from these electrochemical reactions, distinguishing system behavior of living and nonliving systems seems to be as confusing as identifying Bergson's *élan vital.*

A third approach to distinguishing life is to describe living systems by *function or purpose*. This approach is consistent with the scientific method. From this perspective, the scientist might list the characteristics of living systems. By trying to characterize or define living systems by function or purpose and identify characteristics of living systems as distinct from nonliving systems, two difficulties arise. First, it seems a perspective of when the list is complete requires prior knowledge of what living systems are, an unforgivable circularity. Second, it seems impossible to distinguish among living systems. How are living systems like humans different from living systems like trees or viruses?

Maturana developed a different approach, which focused on *organization and pattern* as defining characteristics of living systems. While organization is consistent with a traditional logic, the inclusion of pattern avoids reducing systems to relations, per se, and allows us to concentrate on system-environment interaction and emergence. These notions, as we have seen with Dewey, also are central to the theory of the naturalistic emergence of the mind and have underlying logics of relationship and systems. The notions of pattern and organization are key to understanding and defining living systems and the mind in terms of process and will be explored below.

Organization. First Maturana distinguished between simple and complex systems. Humans, trees, tables, societies, for example, are complex unities composed of many parts. The *organization* of a complex unity or system, living or nonliving, according to Maturana, refers to the *relations among the parts*. Thus, "the relations between components that define a composite unity (system) as a composite unity of that particular kind, constitute its organization" (Maturana, 1980, p. xix). For example, a computer has input devices, some sort of processing unit, and some mechanism for output. How these components are specifically built (that is, the specific structure), varies, but the organization or relation among the component parts remains the same.

By focusing on organization, Maturana's approach represents that of a systems perspective. To avoid reducing systems thinking to mechanics, however, the task is not to find direct relations between parts of a system but to find the superordinate system in which they are connected. Bergson identified that superordinate system as purpose or meaning. Maturana identified emergence and self-creation as the superordinate aspect of living systems. His particular approach to understanding living systems was his exploration of relationship in terms of "autopoiesis" as self-creation, described below.

Autopoiesis. Maturana and Varela (1980) defined the pattern of organization in living systems as *autopoiesis.* Maturana (1980) describes the origin of this term this way:

> [O]ne day, while talking with a friend ... about an essay of his on Don Quixote de La Mancha, in which he analyzed Don Quixote's dilemma of whether to follow the path of arms (*praxis*, action) or the path of letters (*poiesis*, creation, production), and his eventual choice of the path of *praxis* deferring any attempt at *poiesis*, I understood for the first time the power of the word "poiesis" and invented the word we needed: autopoiesis. (p. xvii)

The origin of the word "autopoiesis" emphasizes the notions of self-creation, production, and autonomy. Fleischaker (1990) summarizes Maturana and Varela's notion of autopoiesis by focusing on three characteristics of autopoietic systems. Autopoietic systems are:

1. *closed*, self-bounded, and autonomous;
2. *self-generating* whereby all components of the system are produced by the system; and
3. *self-perpetuating*, namely maintain their identity throughout change.

Boundary, production, and organization both limit and define the historical (and often biological) evolution of an autopoietic system. By shifting focus from defining characteristics of living systems to the process of self-creation through self-perpetuation and self-generation, Maturana has significantly altered the perspective from which living systems can be examined. Living systems are complex, adaptive systems.

As organizationally closed, structurally determined, but interactively open systems, how do living systems interact with their environment? Explaining how living systems interact with their environment is where Maturana's notion of autopoiesis meets Prigogine's ideas about open systems and dissipative structures. Just as with the thermodynamic systems studied by Prigogine, living systems, perturbed by their environment, can react in several ways. Patterns of organization emerge as histories of interactions facilitate organismic growth and small changes become magnified through the recursive process.

The ability for autopoietic systems to interact with their environment is termed by Maturana as *structural plasticity*, the ability of the system to undergo "structural changes as a result of interacting with itself, its environment, or other structurally plastic systems" (Dell, 1985, p. 13), and is a function of being autopoietic. The process of interacting with its environment is *structural coupling*, "recurrent interactions, each of which triggers structural changes in the system" (Capra, 1996, p. 219).

Maturana's biological interpretation of organism-environment interaction is consistent with Dewey's views of coevolution resulting from the organism-environment complex. This is especially apparent as Maturana dismisses environmental causality or stimulus-response chains of interaction in favor of dynamic change that affects both the organism and its environment. Similarly, changes in environmental conditions do not cause changes in the organizational characteristics of a species, as defined by some versions of Darwinism. According to Maturana and Varela (1980), the environment does not *cause* change to occur at all. From the perspective of their structural determinism, causality and control are impossible. The structure of the system determines both the potential for self-generation and the potential for structural coupling with its environment. "The notion of causality is a notion that pertains to the domain of descriptions, and as such it is relevant only in the metadomain" (Maturana, 1980, p. xviii). The psychological experience of causality occurs when "we are able to couple ourselves to objects in such a way that we can bring about a predicted or desired outcome" (Dell, 1985, p. 9). This notion of causality is especially relevant when we consider teacher change, change in classroom practices, and change in student learning in the last chapters of this book.

Thus, relational qualities are built into the very being of autopoietic systems through their structural plasticity while the logic of systems focuses on the organizational qualities and histories of complex evolving structures. As self-organizing, self-moderating, self-referential systems, with the properties of structural plasticity and creativity, living systems adapt through the process of structural coupling. Structural coupling, therefore, is the process whereby autopoietic complexities interact with their environment. "It is the circularity of its organization that makes a living system a unit of interactions, and it is this circularity that it must maintain in order to remain a living system" (Maturana, 1980, p. 9).

Autopoietic unities are complex adaptive systems. They exchange matter and energy, enter into reciprocity relations with their environment and are nonlinear with regard to effects of their interactions. While unpredictable, complex adaptive systems may have interesting and important regularities. These systems at far-from-equilibrium states are sensitive to initial conditions and results of interactions cannot be predicted; nevertheless, there are patterns to their behavior. Hidden and emergent orders are characteristics of nonlinear feedback systems that are of interest to those who study complex adaptive systems.

Both Prigogine and Maturana found emergence in self-organizing structures to involve irreversible processes, as did Prigogine in his studies of systems in far-from-equilibrium states. System complexity and reaction to environmental stresses similarly could not be predictable or traceable to initial conditions. Each system had to be explored and observed relative to its indeterminacy of initial conditions and history of evolution. Patterns of behavior, however, allowed for the study of systems, characterized by the kinds of adaptive behavior in which they engaged. Living systems, in particular, creatively respond in far-from-equilibrium states to achieve new levels of organization and more complex ways of interacting with their environments.

Not Another Grand Narrative

As described above, we must avoid confusing the tools and methods of postmodern science with the meaning structures of postmodern science. The logic of systems is a way of exploring and understanding our experiences and observations and ultimately regarding how living systems interact. Complex adaptive systems theory in particular becomes important as we explore living and social systems. Complex adaptive systems theory, however, must not be confused as representing some hidden truths of how living systems operate and must reflect systems logic to be considered a postmodern theory.

This chapter began with an explanation of the origins of systems logic in systems theory. The *Erwartungshorizonten* of the complexity sciences was traced to systems theory as the underlying systems logic was developed. The middle section of this chapter explored and developed complex adaptive systems as living systems from nonequilibrium physics, dynamical systems theory, and chaos mathematics. Chaos mathematics was explored as a powerful tool for understanding and exploring complexity and will be further discussed in Chapter Seven as social systems are explored. This final section of the chapter will position complex adaptive systems reflecting systems and relational logics in postmodern thought.

Postmodernism and complex adaptive systems

Postmodernism, as described previously, is more a state of affairs or series of events, still tied to modernism, transitioning to the unknown. It is unclear whether during these transitions human and social interactions will become transformed or merely paused, interrupted, or adjusted to accommodate our changing social and natural environments. As described by Latour (1993):

> Postmodernism is a symptom, not a fresh solution. It lives under the modern Constitution, but it no longer believes in the guarantees the (modern) Constitution offers. It senses something has gone awry in the modern critique, but it is not able to do anything but prolong that critique, though without believing in its foundations. (p. 46)

Changes in the sciences have had much to do with the changing perspectives, questions, and challenges to Newtonian ways of thinking and the modernist logic of domination. As described by Fletcher (2000):

> [P]ostmodernists call on us to accept the inescapable partiality of what we know about the world, what we understand about others, and even our perceptions of self. Postmodern theorists have problematized the notion of identity, in particular, arguing that it never comes in an easily defined package and can never be adequately expressed through tidy categories of essentialist attributes.

For postmodernists, the nexus of individual identity and social experience is almost intransigently complex. (p. 28)

Weedon (1997) ties the origins of postmodernism to architecture, also signifying a break from the past and a rejection of the order and permanence previously sought in architecture, as in the sciences. She goes on to suggest that postmodernism defines theories of truth, justice, and inquiry based on the metanarratives of context, subject to change and related to historical development.

Our own logics of relationship and systems provide a perspective of postmodernism that similarly rejects universal truth, recognizes subjectivity, process, and dynamic exchange. The question remains whether the New Sciences, complex adaptive systems theory, and the logics of relationship and systems are just another Grand Narrative, merely replacing the dogmatic perspective of the world as exemplified by Newton with another universal, totalitarian approach to inquiry.

To fully answer this challenge, the third perspective of our postmodern logic needs to be developed. The logic of meaning, together with the logics of relationship and systems, will explore how difference, context, heterogeneity, fragmentation, and indeterminacy are inextricably related to postmodern inquiry, allowing for the historical and evolutionary development of discursive domains and intersubjective inquiry. The next chapter will further explore the logic of meaning as an aspect of postmodern logic.

Notes

1 Fuzzy logic developed out of multivalued logics and was an attempt to create thinking, learning systems that could solve problems and come up with creative solutions. Robots programmed in fuzzy logic, it was hoped, would be able to react in more human ways, making decisions and dealing with ambiguity in ways that machines up to that time could not.

2 Frege attempted to establish the foundations of mathematics in set theory. His early attempts to establish these foundations initiated the field of metamathematics and the study of logical systems.

3 The problems with the foundations of mathematics will be explored more fully in the next chapter. Approaches to the problems of the foundations of mathematics will especially be instructive as parallels are made between the underlying logics of process and systems logic and the logic of domination.

4 The literature refers to Ludwig von Bertalanffy using both his German name, "von Bertalanffy" and his Americanized name "Bertalanffy." I will use the latter.

CHAPTER SIX
The Logic of Meaning

To ask, "What is your aim in philosophy?" and to reply, "To show the fly the way out of the fly-bottle" is ... well, honor where it is due.... But if I were asked to express in one single word what is its most essential feature I would unhesitatingly say: vision.... What is decisive is a new way of seeing and, what goes with it, the will to transform the whole intellectual scene. Wittgenstein[1]

As a classroom teacher, I was always anxious to find puzzles and games for my classroom. The holiday shopping suggestions I gave to my family always included requests for mathematics games and puzzles that demonstrated topological paradoxes. To my nonmathematician family members, it wasn't always obvious what I was asking for. I had to explain that these were the puzzles that usually had ropes and rings and the goal was to remove the ring by manipulating the ropes around the rings and through obstacles. Sometimes these puzzles took the form of sculptures of tangled iron.

Untangling these puzzles requires geometric perspective that recognizes that what appears as a knot is not always a knot. I used to show a mathematics video to my students called "Not Knots" that graphically and pictorially introduces the mathematics of topology with displays of tangled ropes that are often not knots at all! Picturing a knot in a way that you can discover when it is, in fact, not a knot, and exploring the existing classes of "not knots" are related to the study of the branch of mathematics called topology while also illustrative of an entirely new way of seeing these wonderful knot puzzles that were scattered around my room.

Without being aware of the deeper significance of my method, I was enacting Wittgenstein's method of showing the fly out of the fly-bottle.

To change a way of seeing, to visualize the puzzles as topological representations, is to see the problems in a very different way. Seeing the problems as topological knots, the knots disappear and become notknots.

This is exactly the approach to meaning and philosophy that Wittgenstein proposes. In his mind, with a change of aspect or way of seeing, many of our intellectual dilemmas disappear and become nonproblems. "I have found that untying knots in Wittgenstein's remarks unravels the knots in our thinking. The remarks themselves, as he well knew, are the vehicles of change" (Genova, 1995, p. 19). Wittgenstein's approach to showing the fly out of the fly-bottle, his method of inquiry, challenges traditional ideas about language, meaning, and purpose.

The Early Wittgenstein

While Wittgenstein's later work refuted and directly challenged his own early efforts, his early emphasis on action, process, and inquiry pervade all his writings.

> In the early work, (Wittgenstein) speaks of his approach as a kind of Kantian "critique" delineating the boundaries of what can and cannot be said; in the later work, he refers to philosophy as a "therapy" aimed at disabusing people of their philosophical intentions.... Clarification adds a new wrinkle to critique, making it less judgmental, and performance twists therapy into a more demonstrative activity. Philosophers become not poets, critics or therapists for the later Wittgenstein, but performance artists whose only aim is to effect change. (Genova, 1995, p. xv)

Many of the problems with philosophy, according to Wittgenstein in his early work, the *Tractatus (TLP),* result from our illusions about the relationship between how we describe things and the things themselves. For example, he wrote: "The whole modern conception of the world is founded on the illusion that the so-called laws of nature are the explanations of natural phenomena"[2] (*TLP*, 6.371). As discussed in Chapter Two, Galileo's *hypothesis* that mathematics could explore naturally occurring relationships was not intended to suggest that there was a hidden order captured by that very mathematics. However, throughout the Modern Era, mathematics has been (mis)used to represent and confirm the belief that there are underlying laws of nature that are knowable. Similarly, as discussed in Chapter Five, we have confused, in the Modern Era, the tools of science (the mechanical clock, for example), as models to represent some hidden reality. Wittgenstein was responding to these fallacious moves.

Similarly, the use of logic and the derivation of meaning through logical analysis are distinguished by Wittgenstein as a difference between *saying* and *showing*. Logic is the *form* of language and cannot be expressed through it. Logical form only makes its appearance through use. Through the use of language, we "show" or reveal its logic. There is no underlying logic to describe; no way we can get outside of language to discuss its form. The distinction between saying and showing can also be found in Wittgenstein's discussion of proof.

> What makes a proof a proof for Wittgenstein is that it shows the derivation (the connection) of one line from another. In Kantian terms, a proof constructs a concept: "Proof, one might say, does not merely shew (sic.) *that* it is like this, but: *how* it is like this. It shews *how* 13+14 yields 27.... A proof must of course have the character of a model" (*RFM* II, 22).... Again distinguishing between the *that* and the *how,* Wittgenstein this time focuses on the how. (Genova, 1995, pp. 36–37)

The distinction between saying and showing, the *that* and the *how,* in Wittgenstein's early works was also used to develop his ethics. "Ethics" he described, "does not treat the world. Ethics must be a condition of the world, like logic" (Monk, 1990, p. 142). The entirety or wholeness of the world, according to this view, cannot be analyzed and fragmented in an effort to understand the underlying logic or framework. Ethics *is* a condition of the world, revealed through our actions. Metaethics, that is, discussions *about* ethics, are where we run into trouble, according to Wittgenstein.

In his later writings, Wittgenstein develops, extends, and replaces the distinction between saying and showing with an exploration of language games. Wittgenstein's emphasis on changing our ways of seeing through language games, and language games themselves as forms of life, are vital to transforming and escaping the holds of tradition, habit, and culture. "Being aspect-blind makes one incapable of seeing shades of meaning and thus incapable of change" (*PI*, p. 214). Exploring the familiar in new ways requires changing aspect, our ways of seeing, and exploring the language games that entrap us. The logic of meaning, which is Wittgenstein's method, will serve as the third underlying logic of postmodernism. Its relationship to the logics of process and systems will be explored in this chapter as we develop his logic of meaning.

What Is the Meaning of a Word?

Wittgenstein posed the question "What is the meaning of a word?" as the basis of his 1933 and 1934 lectures at Cambridge, which his students compiled to become his *The Blue and Brown Books.* The question of meaning predominated all of Wittgenstein's later works.

More fully developed in his books, published after his death, *Philosophical Investigations* (*PI*) and *On Certainty* (*OC*), Wittgenstein's method of exploring language games delineates an approach to meaning appropriate for a postmodern, connected, holistic, transformative, and evolving theory of meaning and addresses the most fundamental questions of meaning, purpose, and value. The shift from traditional analytic approaches to exploring language represents what has become known as the "linguistic turn" in philosophy.

The linguistic turn

The ancients attempted answers to "*What* is the world made of?" The medievals asked "*Who* made the world?" The moderns querried "*How* does the world work?" Postmodernism asks "*Why*?" (Genova, 1995). Wittgenstein, building on his notion of *Weltbild* and seeing as a way of life, discussed more fully below, shifted focus from external reality to meaning. Answers to "Why" questions reflect ways of seeing that exhibit how we make sense of our world and the meanings we experience.

Wittgenstein's approach to meaning rejected intellectualization or thinking, per se, and focused on seeing, experiencing, and feeling. Key to understanding seeing, experiencing, and feeling, according to Wittgenstein, was language. His language-games approach represents the linguistic turn in philosophy.

Wittgenstein was a key player in introducing and propelling postmodern linguistic philosophy. Linguistic philosophy addresses the problems of meaning resulting from the Cartesian method. While the voices of dissent, including the Romantics, Idealists, and Existentialists in the 18th and 19th centuries, were heard throughout the three hundred year dominance of modernist thinking, their approaches could not overcome modernism's technical rationality.

> On mathematics: Your concept is wrong.—However, I cannot illume the matter by fighting against your words, but only by trying to turn your attention away from certain expressions, illustrations, images, and *towards* the *employment* of the words. (*Z*, 463)

The "linguistic turn" refers to a shift in approach from traditional epistemology to studies of meaning and language use, thereby offering an entirely different method than used by those voices of dissent. As defined by Rorty (1967):

> I shall mean by "linguistic philosophy" the view that philosophical problems are problems which may be solved (or dissolved) either by reforming language, or by understanding more about the language we use. This view is considered by many of its proponents to be the most important philosophical discovery of our time, and, indeed, of the ages. By its opponents, it is interpreted as a sign of the sickness of our souls, a revolt against reason itself, and a self-deceptive attempt (in Russell's phrase) to procure by theft that which one has failed to gain by honest toil. (p. 3)

According to Genova (1995):

> While Wittgenstein was not alone in creating the linguistic turn, he played a central role, especially in Anglo-American philosophy. Under his leadership, philosophy has experienced the first major turning since Kant's Copernican Revolution over two hundred years ago.... By departing from Kant in a different way than does Hegel and all his progeny, Wittgenstein manages a different entry into the Postmodern world. (p. 21)

Wittgenstein's language-games approach neither tries to capture the meaning of words nor provides rules to follow in its method. Rather, this approach unravels the difficulties in philosophy by exploring how meanings and confusions occur. Meaning, Wittgenstein shows, is inextricably connected to context. "One operates with words, and such operations, like moves in a game, are made possible by the context in which the move is made. And to move, to operate, is to act" (Hagberg, 1994, p. 14). "Philosophy is a battle against the bewitchment of our intelligence by means of language" (*PI*, 109).

The meaning of a word and learning how to use it occur within the language game rather than prior to it. There are, from this perspective, no underlying meanings.

The language game is nonreductionistic, contextual, and open to change. An exploration of language games similarly reveals a way of life and an approach to understanding meaning structures from within the game. Finally, the language-games approach describes how we evolve our meaning structures, change our ideas, and fundamentally escape our social, cultural, and personal habits of meaning-making. Each of these components of the language-game approach will be described below.

Language games

Hagberg (1994) describes language games and Wittgenstein's use of them this way:

> Language-games, as employed by Wittgenstein, are internally circumscribed, and are thus autonomous, and they also generate their own range of possible moves without transgressing their own boundaries, and are thus self-sufficient. (pp. 17–18)

The notion of self-sufficiency is important. Sentences have the same sense or meaning, for Wittgenstein, precisely because they are used for the same purposes and understood to mean the same thing, not because of some underlying form or structure of meaning. The same phrase, "five slabs," can mean multiple things, depending on the context, including, for example, an order to bring five slabs or the recording of a number of slabs.

> Imagine a language-game in which A asks and B reports the number of slabs or blocks in a pile, or the colours and shapes of the building-stones that are stacked in such-and-such a place.—Such a report might run: "Five slabs." Now what is the difference between the report or statement "Five slabs" and the order (to bring) "Five slabs!"? Well, it is the part which uttering these words plays in the language-game. (*PI*, 21)

The language game is unique to the situation and therefore is self-sufficient and inextricable from the context. The meaning is not inherent to the words themselves or even the structure of the sentence.

Language-games as a philosophical technique explore the use to which language is put. Thus, rather than answering philosophical questions such as "What is knowledge?" or "What is mathematics?" Wittgenstein used language games to reveal there is no "thing" which answers these questions but "family resemblances" of related and overlapping ways in which we use language. "Game," for example, can refer to board games, games of strategy, games of strength, games played with pieces, games played by people, and so on.

> We are inclined to think that there must be something in common to all games, say, and that this common property is the justification for applying the general term "game" to the various games; whereas games form a *family* the members of which have family likenesses. Some of them have the same nose, others the same eyebrows and others again the same way of walking; and these likenesses overlap. (*Blue Book*, p. 17)

Much of our confusion about collapsing concepts into "things" results from how we use language. That we think of time as an actual entity, for example, is reflected in the language we use. "How much time did it take to get here?" "What is your hourly wage?" "What time is it?" All treat time as a measurable, quantifiable thing. "Telling time," an early concept taught in schools, emphasizes and passes on to our children a mechanical understanding of the quantification of time. It suggests we can point to time the same way we point to a clock.

> Our answer is: the function of the word "now" is entirely different from that of a specification of time—this can easily be seen if we look at the role this word plays in our usage of language, but it is obscured when instead of looking at the *whole language game*, we only look at the contexts, the phrases of language in which the word is used. (*Brown Book*, p. 108)

The language game of time includes all of the ways we discuss time. Our concept of time revealed through these games is the family resemblance of the ways we discuss time in the language games we play.

Wittgenstein's approach to philosophy, therefore, was not to seek truth but to dispel confusion through exploring language games. A key aspect of language games, according to Wittgenstein, is investigating the use of language and complexity of meaning in varying contexts. Meaning, like aesthetics, is often expressed through action rather than words and so language games explore meaning-through-use involving nonverbal as well as verbal behaviors. As described by Monk (1990):

> Rather than trying to answer the traditional questions of aesthetics ("What is beauty?" etc.), Wittgenstein gives a succession of examples to show the artistic appreciation does not consist (as one might think from reading some philosophical discussions of aesthetics) in standing before a painting and saying: "This is beautiful." Appreciation takes a bewildering variety of forms, which differ from culture to culture, and quite often will not consist in *saying* anything. Appreciation will be *shown*, by actions as often as by words, by certain gestures of disgust or satisfaction, by the way we read a work of poetry or play a piece of music, by how often we read or listen to the same piece, and how we do so. These different forms of appreciation do not have any one thing in common that one can isolate in answer to the question: "What is artistic appreciation?" They are, rather, linked by a complicated series of "family resemblances." (p. 405)

In mathematics, proof does not signify truth but establishes meaning through the use of the symbol system employed. As discussed previously, an arithmetic statement like "2 + 2 = 4" is true not by virtue of some underlying reality but is part of the meaning system of arithmetic which is revealed through the language games of mathematics. To get "outside" of these arguments about truth and the relationship between proof and truth, Wittgenstein explores the uses of mathematics through language games.

> The connection between a word and its meaning is to be found, not in a theory, but in a practice, in the *use* of the word. And the direct connection between a rule and its application, between the word and the deed, cannot be elucidated with another rule; it must be *seen:* "Here *seeing* matters essentially: as long as you do not see the new system, you have not got it." (Monk, 1990, p. 308)

As described above, Wittgenstein's approach to and purposes for doing philosophy were to explore language use to dispel the contradictions and paradoxes of our understandings. The logic of meaning is precisely the structure of relationships revealed through language use. Language games, like complex adaptive systems, are autonomous, self-producing, self-sustaining, and self-maintaining. They are also adaptive, evolving, changing, and transforming as contexts change. Language games therefore define their own boundaries and cannot be studied from outside the game itself.

Thus, attempts to understand the meaning of words in an abstract way make no sense from a language-games perspective. To define "five" as used in the sentence "I am purchasing five red apples" cannot be done outside of the context of my purchase. We understand how to use the number five because we have participated in language games, including mathematics, where we have learned when and how to use numbers. To suppose, however, that there is some underlying concept of fiveness that makes these expressions true or that what we mean when we say we are buying five red apples implies some abstract understanding of number, are the sources of many of our philosophical confusions, according to Wittgenstein. "But what is the meaning of the word 'five'? No such thing was in question here, only how the word 'five' is used" (*PI*, 1).

Similarly, to assert the truth of a statement, to state, for example, that such-and-such is true, leads to tangles of meaning. The truth can be revealed through the context, but to say the statement is true requires escaping the context, which is not possible.

Wittgenstein defines valid moves of a language game to be limited by the game itself, which is an aspect of the self-sufficiency of language games. Language games are activities (either through spoken language, gestures, or ritual), according to Wittgenstein, and "to understand a sentence means to understand the language. To understand the language means to be a master of a technique" (*PI*, 199) namely the language game itself as an activity. Put another way:

> For a large class of cases—though not for all—in which we employ the word "meaning" it can be defined thus: the meaning of a word is its use in the language. (*PI*, 43)

The language games in which we participate, in a fundamental way, according to Wittgenstein, are the defining features of the form of our lives. "The limits of language," Wittgenstein described in his earlier work *Tractatus*, "... mean the limits of my world" (*TLP*, 5.62). "To imagine a language means to imagine a form of life" (*PI*, 19).

A form of life

Unlike his own earlier picture-theory of language developed in the *Tractatus*, Wittgenstein, in his later works, distinguishes between language as a representation or projection of meaning and language as a presentational act. Meaning, as exhibited through our actions, through the language games we play, defines the form of our life. "Here the term "language game" is meant to bring into prominence the fact that the *speaking* of a language is part of an activity, or of a form of life" (*PI*, 23).

To explore meanings of language, gesture, and ritual, Wittgenstein invites us to participate in the language games of a culture. One cannot, however, from the outside, understand the meanings without learning the game. The anthropologist studying some indigenous ritual, for example, will not understand its meaning until she can fully participate in the language games of that culture, that is, to partake in their form of life.

The way of life defined by the language games we play similarly reflects a way of seeing, according to Wittgenstein. Wittgenstein distinguishes between "seeing" and "seeing as" with the latter relevant to language games:

> Here also belongs the idea that if you see this leaf as a sample of "leaf shape in general" you *see* it differently from someone who regards it as, say, a sample of this particular shape.... [I]f you *see* the leaf in a particular way, you use it in such–and–such a way or according to such-and-such rules. (*PI*, 74)

"Seeing as" is also relevant to how we see or make sense of our world. Wittgenstein differentiates between worldview (*Weltanschuung*) and world picture (*Weltbild*). The *Weltbild* is a form of life that includes the way we see our world as exhibited through our actions. It is open ended, process and action oriented, and meaning giving. Thinking, seeing, and acting become inextricable through *Weltbild.* Our worldview (*Weltanschuung*), however, is an abstraction, an intellectualization of our *Weltbild.* Once expressed, a worldview becomes our fly-bottle, our trap, the limits of what we can see. "Being aspect-blind makes one incapable of seeing shades of meaning and thus incapable of change" (*PI*, 214). Developing a worldview is to be avoided at all costs, according to Wittgenstein, for once established, it becomes very difficult to escape.

In order to change our way of seeing and transform our *Weltbild*, however, we need to experience the puzzle of our language games, to "see as" in different ways:

> In Kuhn's language, a grammatical movement takes place within a paradigm; it advances the workings of "normal science." The invention of a new way of seeing, on the other hand, is more like the discovery of a new grammar, a new form of life—"And to imagine a language means to imagine a form of life" (*PI*, 19). (Genova, 1995, p. 6)

Imagination is required to change our form of life, to expand or shift our ways of seeing, and to transform our *Weltbild*. Wittgenstein defines it as "changing aspect" or developing a creative sense for "seeing as."

Change of aspect

Wittgenstein does not advance a different way of seeing, only that we must see things in new ways in order to change. "The aspects of things that are most important for us are hidden because of their simplicity and familiarity" (*PI* 129). We need a change of aspect, a change in seeing as. Changing aspect, changing the way we see as, is intimately connected with the language we use and the way we live. At the same time, changing aspect does not change perception but changes both what is seen and how we see it, including how we may respond:

> I contemplate a face, and then suddenly notice its likeness to another. I *see* that it has not changed; and yet I see it differently. I call this experience "noticing an aspect." (*PI* 193)

From my own experience, for example, changing from thinking about my students as things to thinking about them as relationships fundamentally changed how I saw them, how I interacted, and how I interpreted what they did. My students hadn't changed, but how I saw them had. "Once the new possibility is glimpsed, the hold of the previous way of seeing is broken, and we are free to explore familiar terrain in new ways" (Genova, 1995, p. 2).

Change, like meaning and language, occurs in a context. It involves feelings and emotions, not intellectual activity or rationalization. Change is measured by our escape from the fly-bottle, that is, by our untangling the puzzles of our life:

> The expression of a change of aspect is the expression of a *new* perception and at the same time of the perception's being unchanged. (*PI*, p. 196)
>
> Being aspect-blind makes one incapable of seeing shades of meaning and thus incapable of change. (*PI*, p. 214)

Recall, as Rorty (1967) described, the language-games approach provides for a change of aspects as "problems ... may be solved (or dissolved) either by reforming language, or by understanding more about the language we use" (p. 3). Applied to schooling, our educational dilemmas can be approached and perhaps dissolved from a language-games perspective. Changing aspect, seeing things as other than we see them now, may be an important part of transforming public education.

How may our educational problems be dissolved when we consider them from a language-games approach? What new understandings will occur, impacting our actions and values, as we center on meaning? How do the logics of relationship, systems, and meaning together inform and create a new "form of a life," representing a "change in aspect?" While the first two of these questions will be addressed in Chapters Eight and Nine, the following section will offer a way of *seeing* the relationships among systems, relationship, and meaning perspectives *as* a hologram. Changing our metaphors and the language we use to describe what is important to us are crucial to changing our language games of schooling and creating a new *Weltbild* or way of seeing as.

Holographic Metaphor as Seeing As

A hologram is a light wave interference pattern recorded on photographic film. When illuminated, a 3-dimensional image is projected. Dennis Gabor invented and received the Nobel Peace Prize in physics for his work developing holographic techniques in Britain in the late 1940s. His techniques were greatly improved with the invention of lasers in the 1960s. The word "holograph" comes from the Greek words "holos" indicating wholeness or completeness and "gram" meaning message.

In a holograph, every piece contains the information of the whole. The information of each part is "spread out" over the whole object as the light is refracted onto the film. The metaphoric significance of the hologram is seen in psychology (Peat, 1987), theories about how the brain works (Penfield, 1975, Zohar, 1990), cosmology (Briggs & Peat, 1984), theories about death (Ring, 1984), dream theory (Wolf, 1987), and organizations (Wheatley, 1994). The significance of the hologram as a way of seeing–as represents an approach to meaning that will explain the relationships among the logics of relationships, systems, and meanings.

Many aspects of these three perspectives have seemed overlapping. Notions of connectedness, holism, transformation, context, creativity, values, autonomy, meaning, purpose, and care are explored, to varying degrees, in all three perspectives. One could try to argue that these three underlying logics of postmodernism are different perspectives of the same underlying, essential reality yet to be named or discovered. This approach, however, like the debates about the underlying nature of quanta, lead us back into a "thing" world and away from a world of process and relationship.

We could also argue that each of the logics of postmodernism offers three different perspectives or partial views of the same thing, just as walking around a mountain may give us different views of its peaks, elevation, ecology, and erosion. This suggestion is akin to defining the "secret in the middle" discussed previously. Attempts to use the logics of postmodernism to discover some essential "secret in the middle" or underlying truth commit us to the same sins from which Wittgenstein was trying to save us.

The power of the imagery or metaphor of the hologram is that meaning can be seen to pervade all perspectives of the language game of postmodernism. The *Weltbild* of language that uses the holographic metaphor is one where separation and dualisms of subject-object, individual-society, male-female and piece-whole seem to disappear. The construction of holograms even offers suggestions for action. The more diffuse information about the whole is shared among the pieces as a holographic image is being constructed, the clearer the image is. So too in organizations, considered as holograms, can better service be provided to the public as information pervades all aspects of the organization, as the following example describes:

> In a hologram, every part contains enough information, in condensed form, to display the whole.... The image of the whole can be reconstructed from any fragment of the original image. Holograms create wonderful images for the distribution of information in organizations. In fact, we already have an experi-

> ence with organizational holograms in our current approach to customer service. Most organizations acknowledge that when a customer comes in contact with *anyone* from the organization, no matter his or her position, the customer experiences the *total organization*, for good or ill.... We improve customer satisfaction when we recognize and support organizations as holograms ... (and) distribute information broadly across the organization ... (to) strengthen its image. (Wheatley, 1994, pp. 112–113)

Unlike hierarchical arrangements, the holographic relationship entails shared information permeating the entire organization and individuals within the organization. The physicist David Bohm made use of this holographic metaphor to describe the implicate order of the universe (Bohm, 1980). Citing examples in quantum physics and theories of relativity, he pointed out that there seems to be an undivided wholeness to the universe. The relationship between consciousness and the material world, for example, is blurred as quantum relationships are built. How do the "hidden" order and wholeness of the universe reveal an implicate order?

> In the holograph, therefore, the relationship between object and plate is that of the explicate to the implicate and the image is reconstructed by unfolding this implicate order information. The implicate order is also appropriate for the description of society. For example, a crowd of people consists of individuals, each of which has his or her own motivations and beliefs. Nevertheless in certain situations, a football game, a state funeral, a demonstration, a parade, the crowd also has a collective behavior, just as do the electrons in a plasma or the cells in a slime mold slug. In such a case the behavior of the crowd is enfolded in each individual and in turn each individual is unfolded within the whole crowd. (Peat, 1987, pp. 172–173)

While each of our postmodern logics offers a different form of life, a different discourse field, a different way of seeing as, they are all integral to a postmodern meaning system and therefore parts of the holographic language game of postmodernism. Wittgenstein complained of the difficulty of writing about the complexity of his thoughts and the implicate order of his ideas:

> Each of the sentences I write is trying to say the whole thing, i.e. the same thing over and over again; it is as though they were all simply views of one object seen from different angles. (*CV*, p. 7)

The "secret in the middle" is not that there is any "*thing*" that we are describing but that in a very real sense we are trying to describe the whole from within. The postmodern discourse field must develop these discursive perspectives without falling into the trap of the "secret."

We must also take Wittgenstein's warning to heart and avoid introducing postmodern approaches to meaning as *the* way of understanding. "It should be clear why Wittgenstein never bluntly says "I want to change your way of seeing." … Philosophy could only be a catalyst for change" (Genova, 1995, p. 18). Thus, while this chapter and this book are about creating a *Weltbild* or ways of seeing by changing how we talk about schools, Wittgenstein would be terribly disappointed, although probably not particularly surprised, if we used his language-games approach to create our own Grand Narratives or truths. These ideas, these ways of seeing, I hope will serve as catalysts to ways of seeing schooling, teaching, learning, and the curriculum.

One reason Wittgenstein is so hard to read is that he goes to great lengths to avoid doing our thinking for us, to avoid giving us answers to our own problems. I too need to be careful and emphasize that how the logics of relationship, systems, and meaning as ways of seeing as have helped *me.* It is up to each reader to explore the ideas of relationship, systems, and meaning in their daily activities and determine how, perhaps, seeing our school experiences as developing relationships, interconnected wholeness, and meaning systems may dissolve some of our educational confusions and allow us to escape the clutches and confusions of the modernist logic of domination.

Wittgenstein criticizes, however, that a discursive community requires shared language games. Changing the way we talk about schools, essential to changing the institutions themselves, requires cultural participation in the language game of shared meanings. My talking "about" schools from a language-games perspective will have no meaning from the "outside" nor can those who read these ideas engage them meaningfully without participating in the language games of postmodernism:

> "To obey a rule, to make a report, to give an order, to play a game of chess, are customs (uses, institutions)." (*PI*, 199) Thus, meaning has a public character or dimension, which implies a concern for context. (Hagberg, 1994, p. 62)

This is like the dedication to Nietzsche's *Thus Spoke Zarathustra*. "A Book for All and None" implies there is no level of argumentation or rationalization that can convince someone from the "outside" of the rightness of the arguments. Only those who share in the ideas and who choose to see things as described can make sense of the claims. My task is to "show" you a way of seeing as, to offer a change of aspect that may alter how we ultimately talk about schools. Only through changing how we talk about schools can significant change occur.

The next chapter discusses how communication is the medium of meaning that pervades social institutions. In order to understand how a single voice may involve, connect, and influence a larger discursive community, an understanding of social organizations is necessary. The next chapter is the final one in the middle section of my book, the *techné* of postmodernism. How the language games of postmodernism may engage us in changing our ways of seeing schooling will be discussed in the two chapters of the book comprising the third section, the *arté* or creative push to see education from a different *Weltbild*.

Notes

1 As quoted by Genova, 1995, p. 1 from remarks made by Wittgenstein to Stephen Hilmy six weeks before his death.

2 References to these citations will use the footnoting standard more commonly used to refer to Wittgenstein's writings rather than following the reference by publication date and page number. Thus, I will refer to the book by the letters of its name and cite the particular passage number or page number. Specific passages within sections of Wittgenstein's writings will be noted without referencing a specific page number, e.g., *PI*, 4, while passages not within a particular section, most of the second part of the *Philosophical Investigations*, for example, will cite the translator's page number. So, for example, in part II of *PI*, *PI*, p. 218 refers to the page number since no section number is available.

CHAPTER SEVEN
Schools as Learning Organizations

My first year teaching was a delicate balance of meetings, developing the curriculum, learning new names and faces, establishing routines, and putting up with the intercom! It was bad enough that every Tuesday during second period, I had to compete with the coal truck delivering coal outside my window, but the incessant interruptions on the intercom were too much. One day I counted 37 announcements, requests for students to come to the office, and commercials for upcoming student activities. Announcements occurred throughout the class period rather than being saved for the last five minutes. Recognizing that communication in schools is a precious thing, I nevertheless had great disregard for the "squawk box."

This chapter will explore the role of communication and meaning in social systems. Characteristics of social systems that enhance their ability to become learning organizations will be informed by the postmodern logics of relationship, systems, and meaning. Ultimately, seeing schools as learning organizations and helping them develop their adaptive potentials will provide the "change of aspect" discussed by Wittgenstein as necessary for our language games of schooling to move beyond their modernist origins and form a different *Weltbild* or form of life that includes changing the way we see and interact with our world. How, then, may considering schools as learning organizations, and schooling as a social dynamic, provide ways of talking about education and transforming schooling? What is the role of communication and language in these endeavors?

Social Systems

Social organizations have been construed in a variety of ways. The economy, schools, politics, and the law have all been considered structures, collectives, and systems of societies. The problems with talking about social organizations as structures or collectives are precisely those found by Wittgenstein when logic was perceived as underlying structure or a collection of objects with particular characteristics. Logical form is revealed through use. Only through the use of language can we "show" or reveal its logic since there is no underlying logic or meaning to describe. Similarly, social organizations can only be described in reference to their organizational dynamics. Without an underlying logic or makeup, there is no static way to describe the essential features of social organizations such as the economy or public education that captures the complexity of those organizations. Only by observing the dynamics of these organizations can we investigate their meanings. These organizational dynamics may be expressed through systems, meaning, and relational logics.

Social systems as living systems

If meaning is sought through organizational dynamics, then social systems might usefully be considered as living systems. Living systems have at least three defining characteristics: self-generation, autonomy, and organization. Therefore, before a social system is deemed as "living," these three features must be present.

To explore self-generating characteristics, production capabilities of the system need to be identified. Maturana defined autopoietic reproductions as the self-generating characteristics of living systems. Autonomy, the second characteristic of living systems, describes how the system is self- regulating. And self-perpetuating characteristics, how it's identity is maintained through change, needs to be explored, as the third aspect of living systems. Self-generation, self-regulation, and self-perpetuation uniquely characterize any living system by identifying boundary, production, and organizational characteristics of it. How might schools and other social organizations exhibit these properties? We will first consider self-generating potential as a key to the autopoiesis of social systems.

Autopoiesis of Living Systems. To regard social systems as living systems responds to the vision that social systems in some fundamental way have a life of their own, that, in fact, they are themselves capable of adaptive and possibly intelligent behavior. To consider schools or businesses as living systems may allow us to understand and appreciate how the organizational dynamics can be infused with meaning and purpose. Rather than needing to be "managed" or "controlled" by administrators or CEOs, the question that arises is how can we develop organizational capacity for meaning, purpose, and learning. If a living system, schools will exhibit a kind of autonomy, boundaries, and self-referential dynamics that can be uniquely characterized beyond the people who participate in them, allowing for organizational dynamics that contribute to the synergy of relationships within schools. But how might living social systems reproduce themselves?

We will examine three approaches to answering the question of whether social systems are autopoietic and address the question of their reproduction capabilities. In addition, we will see that the reproductions of social systems reveal the underlying logic of our efforts to meaningfully approach our understanding of system dynamics. Viewing social systems as comprised of people, for example, reveals an underlying mechanism consistent with the logic of domination and modernism. Perceiving social systems as meaning systems, as described below, reveals underlying logics of relationship, meaning, and systems.

The first approach to addressing the question of whether social systems are autopoietic places social systems as a subcategory of autopoietic systems but challenges that all autopoietic systems are living systems. This perspective suggests that there are classes of autopoietic systems that cannot be considered living, for example, computer models of autopoietic systems and social systems. As autopoietic but nonliving systems, this position is vague about what the reproduction capabilities of social systems are. Distinguishing between living systems as autopoietic and social systems, as autopoietic but nonliving, is problematic since autopoiesis is a characteristic of living systems, first conceived by Maturana (1980) to define and distinguish living systems from nonliving systems. Without understanding the production capabilities of social systems from this perspective, little is gained in defining social systems as autopoietic but nonliving systems.

The second perspective restricts autopoiesis to organic systems. This means social systems or computer models, for example, by definition, could not be considered living systems. Social systems, from this perspective, could be the medium in which autopoietic systems operate, but are not themselves autopoietic. Social systems, considered as the collection of humans or other living beings (bees, for example), contribute to the autonomy of the individuals by supporting their lives, but are not, from this perspective, living themselves. This position is problematic, however, given that the notion of autopoiesis was developed by Maturana (1980) in the first place to define living systems. Stating that only organic systems can be autopoietic reduces the operational definition of the living to the property of being organic which encounters the problem of distinguishing and delineating properties of living systems (as discussed in Chapter Five), using the scientific method, rather than focusing on processes, as developed by Maturana. It also reveals a mechanistic perspective of social systems, reducing social systems to their component parts. To define living systems by virtue of their being living, as this position basically suggests, is, needless to say, circular and unacceptable.

The third perspective is that all autopoietic systems are living systems but not all autopoietic systems are organic. The focus of this perspective is to examine the boundaries and processes of production of any particular system to determine whether it is autopoietic. If these conditions are met, the system is autopoietic by definition. The question that must be answered from this perspective, therefore, is whether social systems have *boundaries, processes of production, and are self-perpetuating.*

Focusing on the third perspective, therefore, it seems reasonable to ask whether social systems like schools can be autopoietic by examining the definition of autopoiesis. Understanding boundaries, processes of production and self-perpetuating process entails exploration of autopoietic capabilities of a system. Thus we must ask what it means for schools and other social organizations to be autopoietic in these terms, and what difference it might make in understanding schools as autopoietic. By seeing schools as living systems, and exploring how they are capable of self-generating, self-renewing, and self-determining behaviors, we may see schools as adaptive with learning capabilities allowing them to accommodate to the rapid changes of our postmodern times.

Language and meaning in social systems

Mingers (1995) applies Boulding's (1956) classification scheme for complex systems to Maturana and Varela (see Table 7.1) to explain the role language plays in human development and social systems. Building on relationships within the system, levels of complexity are proposed. The role of language will be seen to be crucial in defining social systems as autopoietic. This hierarchy of types of system dynamics is especially useful for understanding relationships among systems as well as relationships within systems at each level of complexity.[1]

The lowest level organizational structures are static, closed systems like bridges and mountains[2] (see Mingers, 1995, pp. 81–82). Closed systems do not exchange energy nor do their structures change. Second-level systems are open, mechanical systems that use energy to continue operations. Clocks and flames on a stove are examples of open mechanical systems that require energy to combat entropy.

Level	Characteristics	Types of Relations	Examples	Domain
Structures & frameworks	Static	Topological constitution	Bridge, mountain	Closed systems
Simple mechanistic	Dynamic, open	Ordered	Solar system, clock, machines	Open systems
Self-regulating systems	Closed-loop control, negative feedback	Specification	Thermostat, body regulation through homeostasis	Dynamical systems
Open, living systems	Self-maintaining, self-producing	Autopoietic	Cell, flame	Dynamical systems
Multicellular living systems	Organized whole, functional parts	Second–order autopoietic, Structural coupling	Plants, fungi	Dynamical systems
Organisms with nervous systems	Closed organization, ability to learn	Interactions with relations	Animals	Linguistic domain
Observing systems	Language, psychology, self-consciousness	Relations of relations	Humans	Consensual domain
Sociocultural systems	Communication roles, values	Third–order autopoiesis, structural coupling between organisms	Families, social colonies, swarms	Social domain

Table 7.1. A hierarchy of complexity (Adapted from Mingers, 1995, pp. 81–82).

Next are self-regulating systems that include thermostats or the body temperature system. These, too, are open systems but, unlike clocks, incorporate negative feedback to regulate themselves. Next, at the fourth level, are simple autopoietic, living systems. The single cell is an example of a self-regulated, self-generating, autopoietic (living) system. Multicellular systems, which provide for structural coupling between cells, such as plants and fungi, are second-order autopoietic systems and occur at the fifth level of Mingers' hierarchy. Sixth-level systems include organisms with nervous systems and are characterized as being capable of interacting with their own internal states. "The state of relative nervous activity becomes itself an object of interaction for the nervous system, leading to further activity" (Mingers, 1995, p. 73). This level falls under the linguistic domain and is comparable to Mead's description of communication through nonverbal gesture (see Mead, 1934/1967). The seventh level is included within the consensual domain, exemplified by humans interacting with relations of relations. Language and self-consciousness occur at this level[3] (described in more detail below). Finally, at the eighth level are social systems where structural coupling occurs between organisms. Families and ant colonies are examples of social systems.

These last three levels of structural complexity of systems distinguish cognizing, living systems from living, noncognizing, and nonliving systems and are characterized by communicative action. "Communication, according to Maturana, is not a transmission of information, but rather a *coordination of behavior* among living organisms through mutual structural coupling" (Capra, 1996, p. 287), a dance of sorts. Learned communication is an example of linguistic behavior and, although not language, serves as the basis for language development. Maturana and Varela (1987) use the example of birdsongs as nonhuman communication among bird-pairs mated for life. These birds communicate, exhibiting linguistic behavior, without using words or language:

> These animals ordinarily live in a dense forest with little or no visual contact. Under these conditions, mating couples form and coordinate through producing a common song.... This melody is peculiar to each couple and is defined during the history of their mating. (Maturana & Varela, 1987, p. 194)

Mead's (1934/1967) example of growling dogs "walking around each other, with tense limbs, bristly hair, and uncovered teeth" (Mead, 1964, p. 111) depicts another example of coordinated action through non-verbal gesture, interpreted within Maturana and Varela's (1987) framework as structural coupling (coordinated action) through (nonlinguistic) communication.

> The act of each dog becomes the stimulus to the other dog for his response. There is then a relationship between these two; and as the act is responded to by the other dog, it, in turn, undergoes change.... We do not assume that the dog says to himself, "If the animal comes from this direction he is going to

> spring at my throat and I will turn in such a way." What does take place is an actual change in his own position due to the direction of the approach of the other dog. (pp. 42–43)

Nonverbal or nonlinguistic communication through gesture implicates meaning and action. The dogs understand their threat to each other and act accordingly. "Communication is realized if and to the extent that understanding (through coordinated action) comes about" (Luhmann, 1995, p. 147).

Language, at the next level of complexity, is the coordination of *linguistic* behaviors. Thus, language is communication (meaning through coordinated action) about communication (focused on linguistic meanings). Capra uses the following example:

> Suppose that every morning my cat meows and runs to the refrigerator. I follow her, take out some milk, and pour it into a bowl, and the cat begins to lap it up. This is communication—a coordination of behavior through recurrent mutual interactions or structural coupling. Now suppose that one morning I don't follow the meowing cat because I know that I've run out of milk. If the cat were somehow able to communicate to me something like "Hey, I've now meowed three times; where is my milk?" that would be language. Her reference to her previous meowing would constitute a communication about communication. (Capra, 1996, p. 289)

Once language occurs, there is the possibility of different levels of linguistic communications reflecting differing levels of abstraction. Thus, language as communication about communications itself becomes a complex system of levels or types of linguistic behaviors. Colors and mathematics are at a different level of abstraction from apples and chairs, for example.

Social structures are the network of actions associated with coordinated communications about communication. Maturana and Varela describe social phenomena as follows:

> We call social phenomena those phenomena that arise in the spontaneous constitution of third-order couplings, and social systems the third-order unities that are thus constituted.... We call *communication* the coordinated behaviors mutually triggered among the members of a social unity. (Maturana & Varela, 1987, p. 193)

Thus, social systems are:

> network(s) of co-ordinates of actions.... As such, a social system is a dynamic system in a continuous flow of changing co-ordinations of actions that remains

> the same as long as these stay contained within the configuration of co-ordinations of actions that defines it as a particular social system. (Maturana, 1988, pp. 67–68)

Language and communication are instrumental in the development of social structures. As "coordinated behavior" shared by members of a social organization, language and communication provide for structural coupling of members of the group and form the social structure as a third-order coupling.[4]

> [H]uman social systems are ... networks of conversations. Accordingly, different human social systems, or societies, differ in the characteristics of the different networks of conversations that constitute them. (Maturana, 1988, p. 68)

Are social structures, as third-order couplings of autopoietic systems, themselves autopoietic? Luhmann argues that social systems can be autopoietic by extending the notions of emergence, organization, and relationship. Quoted by Mingers (1995, p. 141), Luhmann contends

> Social systems use communications as their particular mode of autopoietic reproduction. Their elements are communications which are recursively produced and reproduced by a network of communications and which cannot exist outside such a network. (Luhmann, 1986, p. 174)

Defining communications as the mode of autopoietic reproductions implicates social structures as meaning systems. Luhmann's perspective of social organizations as meaning systems falls within the realm of postmodern logics of relationship, systems, and meanings. From this basis, Luhmann describes the relationship between the openness and closure of social systems to include communications and meaning developing in the context of the social system itself.

This perspective of communicative syntheses also reveals and infuses social systems with a complexity of system/environment relations allowing for interaction among social systems. "A system can communicate not only about itself but also, and perhaps better, about other systems. Unlike life (or biological systems), it has no spatially bounded existence" (Luhmann, 1995, p. 145). Thus, social systems are not as clearly bounded or identifiable as, for example, individual humans. To understand schools as a social system, distinct from yet interconnected with other social institutions like the economy, politics, and the law, makes it hard to determine how to change meaning structures within education without, at the same time, changing meaning dynamics across all domains of our social fabric. "Nevertheless, the system can develop its own boundaries and hold to them because the *reasonableness and unreasonableness* of communication in the system can be constrained" (Luhmann, 1995, p. 145).

By entailing communications as the reproductive medium of social systems, as separate from biological systems, consciousness co-emerges within social structure through meaning relationships. Social systems, as autopoietic systems, in addition to having reproduction capabilities in the form of meaning structures, also have a special kind of autonomous potential that impacts their viability, adaptability, and ability to learn.

> Social systems use communications as their particular mode of autopoietic reproduction. Their elements are communications that are recursively produced and reproduced by a network of communications that cannot exist outside of such a network. Communications are not "living" units, they are not "conscious" units, they are not "actions." Their unity requires a synthesis of three selections, namely information, utterance, and understanding (including misunderstanding). This synthesis is produced by the network of communication, not by some kind of inherent power of consciousness, nor by the inherent quality of the information. (Luhmann, 1990, p. 3)

Social systems, seen as living systems with meaning as their reproductive capability, implicate logics or meaning, relationship, and systems. Another aspect of living systems is their ability to define their own borders. Living systems are in open dynamics with their environment while also maintaining a closure that allows for creative emergence and reorganization. If social systems are autopoietic by virtue of their meaning structures, how do they exhibit boundary conditions? If schools, for example, are complex meaning structures, which are part of the social environment, how are schools distinguished from banks, businesses, or factories?

Socioautonomy

Just as systems logic compels us to explore whether a social system is autopoietic, relational logic suggests the need to explore how a system relates to itself and its environment. If it has the capacity to self-regulate, the system is said to be autonomous.

For individuals, the common notion of autonomy as "freedom" ignores the complexity of relationships between individual humans and the environment. Within social constraints, even within a prison, for example, an individual may exhibit autonomous behavior. For humans as cognizing subjects, autonomy, as the ability to self-regulate, entails creative problem-solving and decision-making even within these constraints.

Piaget (1965) suggested human development occurs on a continuum from anomy (nonregulation) to heteronomy (regulation by others) to autonomy (self-regulation). Social context, including unequal power relations, may prevent or inhibit the development of autonomy. Cooperation supports individual growth and the development of autonomy through reciprocal interaction. Cooperation, according to Piaget, entails "striving to attain a common goal while coordinating one's own feelings and perspectives with a consciousness of another's feelings and perspectives" (DeVries, 1997, p. 5). The coordination of activity balancing power, conflict, and authority with cooperation, exploration, and choice is essential to the development of autonomy.

Social systems also may develop autonomy through their own kind of self-actualizing processes. Socioautonomy was originally defined as, *"the creative potential for self-creation and self-production of social systems"* (Fleener & Rodgers, 1999). As in the development of individual autonomy, there seems to be a progression from anomy to heteronomy to autonomy in social systems. As meaning systems, two key features of socioautonomy are shared meanings and communication. Cooperation and conflict help a social system develop its identity and purpose while meaning relationships and communication characterize and distinguish it from other social systems.

The third aspect of an autopoietic system is how the system becomes self-actuating. Through autopoietic reproductions and autonomous action, living systems interact with their environments and with their own internal states. How do they maintain their identity throughout this process? Humans, for example, regenerate our cells every day. Over a two–year period, we are comprised of mostly new cells, yet we somehow maintain our identity throughout this process. How do social organizations maintain identity separate from their environment?

Identity through interactions

Luhmann's social systems theory, Wittgenstein's language-games approach, Maturana's notion of autopoiesis, and Piaget's notion of autonomy in social situations ground systems, meaning, and process approaches to understanding social structures as living systems. Language games are the enactment of social dynamics and thus are the existing meaning structures in specific social settings. At the same time, language games reveal the dynamics among authority, autonomy, and the construction of shared meanings as socioautonomy emerges. Communication evolves as meaning emerges. Communications, as the reproductive capability of social systems, entail meaning as constitutive of social systems and, through the language games that constitute their organization, maintain the identity of the organization as the language games evolve.

Not all social systems may develop socioautonomy, the capacity for autopoietic reproductions of meanings that become infused with purpose and value. How do social systems, as meaning systems, become learning organizations with self-actualizing potential and fully autopoietic capabilities? How can schools be seen as complex meaning structures capable of adaptive, self-regulating, self-generating, and self-renewing characteristics? How can schools, viewed as learning organizations, transform themselves, evolve meanings, and escape from their own constraints of repeated failures? While this last question will be more fully addressed in Chapter Nine, how business and industry have integrated learning organizational dynamics will provide a perspective for how we might understand schools as learning organizations.

Learning Organizations

Business and industry have already explored the adaptive capabilities of organizations, reconceptualizing organizational structures and challenging factory-model notions of leadership and relationship. These same approaches are now accessible to and important for revisioning and developing the potential of schools as learning organizations.

Senge (1990) explored leadership and organizational change in business from a learning organization perspective. He described the basic properties of a learning organization to be "an organization that is continually expanding its capacity to create its future.... [F]or a learning organization, "adaptive learning" must be joined by "generative learning," learning that enhances our capacity to create" (p. 14). Kofman and Senge (1995) are careful, however, not to treat learning organizations as "things" but recognize that their importance is in the meanings offered by learning organization language.

> "Learning organization" is a category that we create in language. Like every linguistic creation, this category is a double-edged sword that can be empowering or tranquilizing. The difference lies in whether we see language as a set of labels that describe a preexisting reality, or as a medium in which we can articulate new models for living together. (p. 32)

The work of Brian Arthur at the Santa Fe Institute further supports exploring social systems as learning organizations with the ability to adapt and transform (see Waldrop, 1993). Leadership and management books supporting the principles of systems thinking and organizational dynamics have proliferated in the last decade of the 20th century, especially as "dot.com" businesses have dramatically changed the landscape of global markets. Learning organization ideas are especially appropriate in "information rich" organizations such as schools and technology industries. In economic theory, for example, information technology markets infused with the ability to learn have a tremendous advantage over those that do not.

> The parts of the economy that are knowledge-based ... are largely subject to increasing returns. Products such as computers, pharmaceuticals, missiles, aircraft, automobiles, software, telecommunications equipment or fiber optics are complicated to design and to manufacture. They require large initial investments in research, development and tooling, but once sales begin, incremental production is relatively cheap.... As more units are built, unit costs continue to fall and profits increase.... Not only do the costs of producing high-technology products fall as a company makes more of them, but the benefits of using them increase. (Arthur, 1990, p. 93)

In such industries, built-in adaptive characteristics prevent the need to start over every time the market changes and allow rapid responses to consumer demand and technological advances.

Stacey's book *Managing the Unknowable* (1992) describes how strategic positioning of organizations "at the edge of chaos" represents a paradigm shift from old approaches of control and stasis. Organizational viability is not determined by "running the course" but by the ability to change, and system structures that encourage and facilitate adaptive behavior and leadership at all levels of the organization are vital. These features of the learning organization will be considered as we re-envision schooling from the perspective of the learning organization.

Wittgenstein's notion of "seeing as" is an important aspect of changing how we look at schools and social organizations as autopoietic, living systems. Implicit in his "change of aspect" is the role of language. Language and language games are the meaning structures of our social fabric. Changing how we see schools as learning organizations in a very real sense changes the language games of schooling and the meaning dynamics of schools.

Bethanis (1995) discusses the role of metaphor in meaning systems from three perspectives. *Root metaphors* "are assumptions that are deep in our memory or on the surface. In either case, our assumptions about anything … come forth when we speak" (p. 190). The clockwork universe is a root metaphor for describing how parts interact, how we can understand the inner workings of something, and how we can become master clockmakers through technical control. Other root metaphors, often not so obvious, are hidden in the words we use that reveal a logic of control, domination, hierarchy, and mechanism. "Time-on-task," "classroom management," and "standards" are all expressions used in schooling that reveal meaning structures with root metaphors of mechanism and control. Through technical control we believe we can manage time, behavior, or academic performance, for example.

Bridging metaphors are valuable in reaching mutual understandings and shared meanings. For example, when describing how to plot points on the Cartesian plane, an algebra teacher may describe to her students how Descartes first noticed that the position of a fly on the ceiling could be uniquely determined by counting the number of tiles down and the number of tiles over along intersecting walls. The teacher might thus use Descartes' discovery of such mappings as a bridge for understanding current uses and meanings of the Cartesian plane. Students create their own bridging metaphors when, for example, after playing the game "Battleship," a student finds new meaning and understanding for plotting points on the Cartesian plane in algebra class. Bridging metaphors are also useful in describing emotions and feelings of pain and discomfort. "It's like someone is rolling a pin-cushion up and down my leg" is a bridging metaphor one might use to describe a pain to the doctor.

Root and bridging metaphors are integral to our language games as we negotiate and share meanings. *Generative metaphors*, however, extend meanings and evolve our language games. They "provide the impetus for conversants to disclose new understandings … for new meanings to emerge" (Bethanis, 1995, p. 191). Waldrop (1993) describes how quantum physics and New Science have provided generative metaphors in the emergence of many new fields of study. Brian Arthur's work in economics, for example, has dramatically affected how we perceive organizational viability in the emerging marketplace.

> Neoclassical theory assumes that the economy is entirely dominated by *negative* feedback: the tendency of small effects to die away.… The dying-away tendency was implicit in the economic doctrine of "diminishing returns."… Negative feedback/diminishing returns is what underlies the whole neoclassical vision of harmony, stability, and equilibrium in the economy. But, … what happens if you have *increasing* returns?… Positive feedback, increasing returns—maybe these things did happen in the real economy. Maybe they ex-

> plained the liveliness, the complexity, the richness he saw in the real-world economy all around him. (pp. 34–35)

Arthur's rejection of the root metaphors inherent in the law of diminishing returns and shift to seeing the economy as a complex adaptive system with viability inherent in the law of "increasing returns" and nonlinearity has completely changed ideas about the economy and organizational viability in the changing economy. Rather than ignoring small changes and working toward stability, the economy as a complex adaptive system is seen to require and thrive on diversity and volatility with the maximum gains occurring "on the edge of chaos." Many economists rejected these notions and refused to consider the implications of an economy driven by the law of increasing returns, however. Traditional ideas about the economy emphasize security, predictability, equilibrium, and control rather than creativity and nonequilibrium. Lacking control and predictability, operating at the "edge of chaos" makes managing such an economy impossible. The economy, seen as a complex adaptive system with potential for creativity, unpredictability, and self-organizing capabilities, does not fit the language and paradigms of traditional economic theory.

The generative metaphors of learning organizational dynamics are found in our logics of systems, meanings, and relationship. Complex adaptive systems, relational dynamics, and language games as meaning structures are different ways of seeing organizational viability within social systems. Developing, sustaining, and growing a learning organization with adaptive potential requires the diffusion of meaning structures throughout the organization. Developing organizational capacity from a complex systems perspective, therefore, requires nurturing and capitalizing on the creative, self-generating, self-organizing properties of socio-autonomous, autopoietic meaning systems.

Changing how we see organizational processes, however, entails changing the language we use to describe how organizations operate, what they are for, and the role of meaning within them. Stacey (1992), for example, describes how we must change our ideas about control and power. "Managers" should not assign tasks and dictate quality-control measures. Rather, from a learning organization perspective, meaning and purpose should pervade all aspects of the organization. Risk-taking, creativity, multiple perspectives and approaches to problems, and learning should be encouraged rather than squelched from his perspective of leadership in a learning organization.

> The key message of the dynamic systems model is that a preoccupation with order, stability, and consistency in all time frames damages management's creativity and ability to cope with the unknowable.... The key question managers face, then, is not how to maintain stable equilibrium but how to establish sufficient constrained instability to provoke complex learning. (Stacey, 1992, pp. 202–203)

Rethinking problems as "communicative obstacles" or "barriers to creativity," for example, completely changes how we perceive our predicaments. Rather than focusing on products, individuals, and control, understanding and appreciating the dynamics of relationship, implicit meaning, and system characteristics can be reflected by changing how we talk about organizations. The generative metaphors of relationship, systems, and meaning logics offer a language of process and complexity that sustain and contribute to ideas about organizational viability and the capacity to adapt to social changes at an organizational level.

Hargreaves (1994) and Fullan (1993) describe their work with teachers and schools from the perspectives of learning organizations. The challenges of re–envisioning schools as learning organizations are to 1) advance the learning capability of all schools, 2) culturally and structurally embed adaptive abilities of schools, and 3) pervade meaning and purpose at the heart of schools. These can be seen through the generative abilities of the language games of schooling.

Unlike that classroom from so many years ago described at the beginning of this chapter, communication-rich learning organizations are not about "announcements over an intercom," but are meaning systems where language games reflect purpose and value at all dimensions of organizational dynamics. Pronouncements from central administration or mandates from key management personnel prevent the flow of communication upon which learning organizations thrive. Tapping the creative potential of individuals within an organization entails evolving social dynamics and the emergence of socioautonomy along with individual autonomy.

The next chapter will discuss the role of the curriculum to change our language games of schooling, reflecting the inherent purpose and meaning of schools as learning organizations, and the implications of schools as learning organizations for social and individual meaning. Our "change of aspect" reflected in seeing schools as learning organizations entails the need for our language games of schooling to move beyond our modernist origins and form a different *Weltbild* or form of life reflected in how we talk about the meaning and purpose of schooling.

Section Summary

These last four chapters comprising the techniques of postmodern logics provide the language, metaphors, and ways of talking about schooling to overcome our modernist tendencies and to provide a way of seeing as and thinking about the world imbued with meaning and purpose. Like a hologram, together these postmodern logics provide an image of how schools might be. Also like a hologram, the perspectives of relationship, meaning, and systems are not mutually exclusive. The clarity of our holographic image of the postmodern world depends on the interconnectedness of relationship, meaning, and systems perspectives. These are the lenses through which we will focus beyond the threshold of modernism where we now stand. The final section of the book will explore curriculum, instruction, and learning, informed by the logics of relationship, systems, and meanings from within the dynamics of learning organizations.

Notes

[1] Although Mingers (1995) represents the relationship among systems as a hierarchy, it could perhaps more meaningfully be explored as a recursive and reflective continuum, better capturing not the separation among system types, but their evolutionary and relational characteristics. Although I will use the descriptions and hierarchy he describes, I will also insert linking language to better relate the kinds of systems discussed.

[2] Mingers (1995) uses mountains as an example of static structures although there is a great deal of debate about this. While the position that suggests mountains are living, dynamic structures better reflects the kind of dynamics discussed in this book, I have left mountains as an example of a static structure since both Mingers and many individuals tend to experience them this way.

[3] While Mead and many others identify language as a distinguishing characteristic of humans, I do not hold that position. I do not think we can nor should we assume we can determine the communication and reflective abilities of other animals. There is evidence, especially among other mammals, that language and reflective abilities do occur.

[4] This definition of social systems (I believe correctly) implicates language as a characteristic of nonhuman animals as well.

PART III: CREATION

Overture

CHAPTER EIGHT
Curriculum Dynamics

Synthesizing the ideas of this book without reducing them to prescriptions for action or presenting them as ultimate answers to our problems is a challenge. Wittgenstein also was concerned that the spirit of his ideas would become reduced and interpreted or used to prevent thinking rather than encourage it. He described his method and concerns this way:

> For more than one reason what I publish here will have points of contact with what other people are writing to-day.... I make them public with doubtful feelings. It is not impossible that it should fall to the lot of this work, in its poverty and in the darkness of this time, to bring light into one brain or another—but, of course, it is not likely. I should not like my writing to spare other people the trouble of thinking. But, if possible, to stimulate someone to thoughts of his own. (Wittgenstein, *PI*, p. vi)

In these final chapters, as I draw some tentative conclusions of the potential impact of postmodern logics on our ways of seeing and forms of life, I too am hopeful and somewhat more optimistic that my metaphors and visions will serve as prompts for generative discourse rather than as complete answers to our troubles in education. In fact, as Wittgenstein discussed with his language-games approach, what we should find is not solutions to our problems, but dissolutions of them. Challenging our old games and inventing new ones, however, will create new problems to be addressed. New challenges and opportunities face us in education as we look to transform our language games of schooling. The process of identifying and exploring these new problems should re-focus our attention away from our past difficulties as our language games provide new meanings, purposes, and offer hope to meet new challenges by constantly re-envisioning our dilemmas and responses, enabling us to escape paralysis of thought and action.

Many of the metaphors of quantum physics, complex adaptive systems, chaos, and complexity will serve as "ways of seeing as." Their use should be interpreted and understood for their metaphoric power, and not as underlying principles or statements of theory. While comparisons with other perspectives of postmodern curriculum will be offered only sparingly, it is my intent that the logics of postmodernism will engage others in ways that lead to evolving meanings, purposes, and values in education. Especially to my students, classroom teachers, administrators, colleagues, and political leaders, I hope these ideas gain relevance in their lives and impact existing meaning structures comprising our social and institutional fabric.

This chapter will discuss the role of the curriculum in changing (or perpetuating) our language games of schooling. The curriculum reflects the inherent purpose and meaning of schools. We will explore the curriculum as emergent, adaptive, relational and meaningful and its possible impact, from this perspective, the transforming of schooling. The logics of meaning, relationship, and systems are the lenses through which we will envision the holographic image of schooling infused by meaning and purpose through curriculum dynamics. The curriculum we envision will be seen to support and offer the meanings by which schools can become learning organizations, allowing us to transcend the logic of domination and our Western modernist limitations, biases, and entrapments. We will also see, however, how traditional debates about the curriculum and the traditional language games of schooling doom us to repeat the curriculum cycles we have seen over the past 100 years. Curriculum perceived as a "thing" rather than as a dynamic process, ensnares us in the same old debates and language games of modernism.

Curriculum as Language Games

Curriculum both defines and is the language game of schooling. If perceived as a "thing"—a set of lesson plans, goals, standards, or outcomes—the language games reveal an underlying logic of domination. To master the standards, achieve our goals, or have positive outcomes, requires technical control and mastery. Implicit in this perspective is the need for managers in the form of teachers, parents, principals, school boards, to ensure good quality, high standards, and above-average achievement for all children. Above-average achievement for all children, of course, is mathematically impossible but nevertheless is often stated as one of the goals of modern schooling.

Perceived as a dynamic, however, we change the language games of the curriculum and the implicit meaning structures of schooling. Like quanta as complex relationships, the curriculum, viewed as a self-organizing process, entails entirely different ways of understanding and organizing our interactions. The way of envisioning the curriculum described below as a dynamic process and complex of relationships reflects the underlying logics of relationship, systems, and meaning.

Self-organizing curriculum

Doll (1993) predicts, "if a post-modern pedagogy is to emerge, ... it will center around the concept of self-organization" (p. 163). Self-organization occurs not when there is control but when there is anomaly, perturbation, difficulty, and complexity. Rather than the curriculum limiting possibilities and keeping all students on the same track, a postmodern curriculum, from this perspective, should encourage diversity, multiple perspectives, and exploration.

> Curriculum will be viewed not as a set, a priori "course to be run," but as a passage of personal transformation. This change of focus and subject will place more emphasis on the runner running and on the patterns emerging as many runners run, and less emphasis on the course run, although neither the runners nor the course can be dichotomously split. Organization and transformation will emerge from the activity itself, not be set prior to the activity. (p. 4)

Doll suggests a curriculum matrix, web-like and complex, to characterize and reflect the complexity of organization and emergence of pattern envisioned in a postmodern curriculum. A matrix, like a web, has no beginning or end, and while it is bounded and structured, these features of its organization emerge through construction and expand through use. A web is also nonlinear and nonsequential yet has points of intersection, connection, and focus. A web can grow from simple origins to be more connected, complex, and coherent. The curriculum itself grows not from the known, but to the unknown, changing the participants as well as the educational landscape as it goes.

> Curriculum is a process—not of transmitting what is (absolutely) known but of exploring what is unknown; and through exploration students and teachers "clear the land" together, thereby transforming both the land and themselves. (p. 155)

Doll (1993) suggests four ways of exploring the curriculum as a self-organizing process. Such a curriculum should be *rich, recursive, relational, and rigorous.* The language games of these meanings change from a postmodern perspective. Interpreting the meanings of these ideas requires not only changing how we talk about the curriculum, but also changing the images or metaphors these ideas represent.

Curriculum matrix

The *richness* of a postmodern curriculum "refers to a curriculum's depth, to its layers of meaning, to its multiple possibilities or interpretations" (Doll, 1993, p. 176). The images we may have associated with "depth" and "layers," from a modernist perspective, may include digging holes or peeling an onion. In digging a hole, we are digging deeper, going to greater depths through hard work. We hear similar words all the time, especially in sports. "Dig deeper" implies trying harder, working harder, and tapping more of your inner strength and resources. Similarly, with onions, if we think of the curriculum as an onion, we only scratch the surface when we present topics superficially. Going to deeper levels might imply a more thorough and complete investigation of some particular concept, including making connections with other ideas. The phrase "a mile wide and an inch deep" has been used to refer to the lack of depth of the American mathematics curriculum.

Neither of these metaphors captures Doll's (1993) meaning of *richness*, however, because both take as given content knowledge and assume shared understandings of what we think is important for students to know in any particular field of study. Content as the curriculum implicates curriculum perceived as a "thing." Both perspectives, digging deeper and going to deeper layers, are still treating the curriculum as a given "thing" rather than as a process.

I think the Mandelbrot set can serve as a powerful metaphor for understanding Doll's (1993) notion of a rich curriculum that captures the process and emerging relationships of a postmodern curriculum. As we move around the edges of the Mandelbrot set, we see infinite complexity, while at the same time common patterns and relationships continue to appear. There are computer programs, many now found on the Internet, that allow us to traverse the edges of the Mandelbrot set, experiencing its complexity and the relationships among the patterns that emerge.[1] This type of "depth" offers "the right amount of indeterminacy, anomaly, inefficiency, chaos, disequilibrium, dissipation, lived experience" (Doll, 1993, p. 176) that Doll characterizes as important for a rich curriculum. For example, as we move around the edges of the Mandelbrot set, we may notice the reappearance of the Julia rabbit or the Mandelbrot bulb popping up out of images that seem unconnected or novel. (See Figures 8.1a-d.) Beneath what appears to be chaos are underlying orders that seem to re-emerge just as we thought we were in unknown territories. This kind of complexity is more suggestive of the interconnectedness and depth associated with relational meanings and comprehensive explorations than a hole that eventually may entrap us. The rich curriculum, seen as a Mandelbrot set, maintains these characteristic forms while providing for infinite variability, emergence, and exploration at deeper and deeper levels.

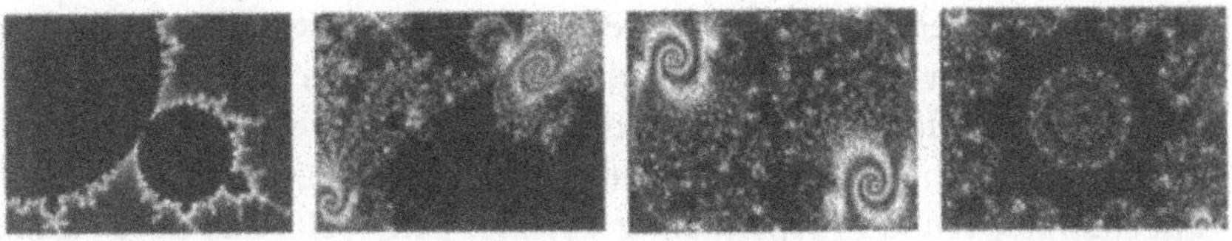

Figures 8.1a–d. Perspectives of the Mandelbrot set (courtesy of Eric Schol[2])

Moving about its edges, along the borders of chaos and order, we can also look at a specific region of the Mandelbrot set and explore it for interesting and emerging properties. Thus, for example, when we look at the Mandelbrot set, we notice "bulbs" of varying sizes attached to the set. When examining these different regions of the Mandelbrot set more closely, we learn that the different bulbs of the set have different characteristics. The three bulb is different from the five bulb, and even the emerging baby Mandelbrot sets that appear as regions along the edges of the three bulb are different from those attached to the five bulb.[3] These are like the different content areas studied in school and comprising the curriculum. As we experience meaning, functionality, purpose, and value in the different content areas, so we experience the unique characteristics found in each of the Mandelbrot regions. The different bulbs of the Mandelbrot set, like the different content areas, are similar, and related, yet different and unique.

To study each content area piecemeal and not recognize its relationship with other domains of meaning, however, just as to study each bulb in the Mandelbrot set without understanding how it relates to the entire set, is to miss an important aspect of the overall meaning and purpose of the curriculum, on the one hand, and the relational characteristics of self-similarity and internal consistency in the Mandelbrot set, on the other. Between the two and three bulb is the five bulb, and between the three and five bulb is the eight bulb. These relationships are significant and lend mystery and beauty to the internal dynamics of the Mandelbrot set taken as a whole. By ignoring these relationships, by only exploring individual bulbs, we fail to see these relationships or important characteristics of the entire whole. This is what happens in schools as we become too insulated, fragmenting the curriculum into content areas, and forgetting that richness is missed when we fail to explore the relationships among different domains of study.

Recursion is a second aspect of Doll's curriculum matrix. In a traditional sense, recursion is often treated as repetition or cycling. Recursion is more than repetition or recycling, however, but is a powerful component of growth. In biological systems, for example, recursion is important as cells multiply and differentiate. Recursion includes an element of reflection not found in repetition or cycling. Only cells that have "gone mad" replicate without reflection. Cancer cells, for example, repeat their unexamined growth, failing to recognize when to stop. When you cut yourself, the cells grow rapidly across the wound to make the repair. If, however, they fail to recognize when the repair has been made, failing to recognize that the two sides of the wound have now been patched over, they continue to grow outside the bounds of the wound. Critical to recursion, therefore, is a reflective, evaluative component. Repetition without decision-making or consideration continues out of control, like cancer cells that lack contact inhibition and continue to replicate until eventually they must be cut out or the source that supports them dies. With recursion there is a creative, evaluative component.

One of my early teaching experiences included judging a science fair contest. I was assigned the computer programming entries because of my expertise in that area. First prize that year was awarded to the ninth grader who wrote an elegant program that solved the Tower of Hanoi problem using recursion. According to ancient lore, the world will come to an end when the monks at the Tower remove all 64 disks from one stake and place them in correct order on another stake. (See Figure 8.2 with four disks.) The rules for moving the disks are simple: Only one disk may be moved at a time. A larger disk may not be placed on a smaller disk. Disks, one at a time, may be placed on any stake. With these simple rules, the solution to the problem using recursion is elegant, requiring only a handful of steps. Without recursion, a series of if-then statements must be processed and reused for as many disks as you may have.[4] The "brute force" method of solving this problem fails to capture the creative, recursive nature of the problem. Like the child who can only deal in the concrete, the patterns of decision-making and arrangements of the disks are lost when we only consider each move as a separate act. The decision-making component of the process includes solving a small problem (where to move the next disk) and repeating the process, with checks about the validity of moves.

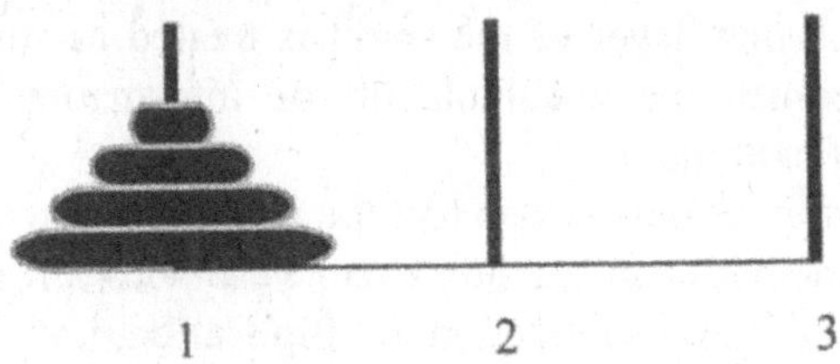

Figure 8.2. Tower of Hanoi puzzle

The beauty and power of the recursive solution is its unwrapping and dynamic relationship among "possible worlds." In trying to explain recursion to my programming students, I used to take a piece of notebook paper that had simple instructions for solving a problem (like moving a disk from one tower to another). Each time a program recurses, it doesn't just go back to the beginning; it creates a new world where the initial conditions are different from the previous world and where some evaluation occurs or a decision is made. As the recursive process continues, new worlds are created each time (in the middle of one world) new possibilities are created. I represented this to my students by starting a new page in the notebook for each possible world. Once we are finished with one possible world, that is, once we reach the end of that world, we leave that page to return not to the original page but to return to the world that sent us to the one we just completed. For example, once we have made a valid move in one possible world in the recursive program of the tower problem, we complete that world and return to the previous world, the second to the last page in our possible worlds notebook. It is the unwrapping of the recursion, the falling out of possibilities, that gives recursion its power as we traverse, backwards, through the notebook, through our possible worlds. The efficiency of the recursive process is that we do not need to remember where we left off or where we came from. The recursion wraps and unwraps itself once we have come up with the overall decision patterns and valid moves.

Compare this power of recursion to the complexity of repetition. "Spiraled" curriculum without reflection and creation is like solving the tower problem by brute force. As more and more disks are added, we need to lengthen our instructions to solve the problem. (For example, after three disks have been moved to stake two and stacked appropriately, we might go through the following steps: "Now take largest disk 4 and put it on the empty stake 3. Then move disk 1 on top of disk 4 and then put 2 back to the now empty stake 1. Now put disk 1 top of disk 2, move 3 on top of 4, 1 back to the now empty stake 2, 2 on top of 3, and 1 on top of 2," completing the puzzle.) Spiraling may revisit old ideas in new ways, but each new layer of the spiral is treated as something discrete. Learning becomes the accumulation of information rather than the exploration of patterns.

Repetition in schools is also like the cancer that grows out of control. To begin mathematics instruction every year with a review of the basic operations of addition, subtraction, multiplication, and division of whole numbers in upper elementary, and of fractions in middle school, is mind numbing for most children. Eventually this repetition either kills its source, or becomes so unwieldy that nothing seems to be able to contain it and no new information can be gleaned. When through repetition our curriculum becomes too unwieldy, just as in oncology when the cancer grows out of control, our focus shifts from what is good for and interesting to each child to how to control the growth and manage the curriculum. We become concerned with managing the disease we have created rather than helping children learn.

The roles of "practice" and repetition in schools have long traditions. Parents are concerned when their second graders are not given mathematics facts to practice at home. Flash cards and other memory devices are purchased to support the process. Although for some students these methods may be effective, it is probably not because of the repetition itself but because the students are engaged in reflection and creativity. They are experiencing these tasks recursively and mindfully, rather than repetitively, with reflection and connections occurring throughout the process. While one student memorizes 7 + 6 = 13 after much repetition, another starts to see 7 + 6 "is like" 6 + 6 plus one more or "6 + 4 (is 10) and three more" in another context. They both respond with "13" but one has developed a sense of numbers while the other has memorized a discrete, inert fact. When the context changes to modular arithmetic, for example telling time on a clock, the first child may have difficulty recognizing that 7 p.m. plus 6 hours is 1 a.m., while the one who has seen ways of "playing with the numbers" recognizes this relationship quite easily.

In generating fractals, we confuse the recursive process with repetition. The code for creating the Sierpinsky triangle, for example, is quite short. To "grow" this fractal, there are again some very simple rules. Pick a starting point. Pick a corner. Move halfway between the last point and the corner just picked, and place the next point there. Repeat this process. Here the recursion includes a decision on how to pick a corner. This simple sequence of decisions and moves, as easy as using a random number generator to pick one of the three corners and the recursive process of doing this over and over again, creates the Sierpinsky triangle. Characteristic of all fractals, the Sierpinsky triangle exhibits self-similarity across scales, emergent pattern, and fractal dimension. Each Sierpinsky triangle looks the same, yet the path to create each triangle is unique. For example, in creating a particular Sierpinsky triangle, we may start with the point at the center of the triangle and move halfway between there and vertex A, then halfway between that point and B, then again between that last point and A again, and so on. Although this path is unique and the likelihood of any two Sierpinsky triangles being built following the exact same path is extremely unlikely, the same pattern emerges each time: triangles within triangles within triangles, and so on.

Figure 8.3. Sierpinsky triangle

Although there are patterns of relationships in common experiences, each individual student, as the complex of many recursive, reflective, relational processes, is unique. In an emergent, postmodern curriculum, we are interested in both exploring these individual paths as well as understanding these emerging patterns.

The Sierpinsky triangle is also a good visual metaphor for Doll's (1993) third aspect of a relational, emergent curriculum: *relations*. One perspective of relations in the curriculum reveals the connections among the emergent patterns of the curriculum process. Here the curriculum might be conceived as the actions and products of these interactions and evolutions as it unfolds, the "path created in the running" rather than the racecourse itself. This perspective of the curriculum is traditionally perceived as the "hidden" curriculum, emergent curriculum, or "currere." Relations in the curriculum, therefore, from this interpretation, are the internal and evolving processes of classroom interactions and unfolding of experiences rather than the planned curriculum or curriculum guides sitting on shelves and developed by the experts. This perspective of curriculum relations emphasizes the process rather than treating the curriculum as a product or "thing."

The emerging patterns of the curriculum viewed in relationship, however, are not necessarily discrete, forming a pathway of interactions. The path may be broken or disconnected, and include spontaneity and unpredictability. In our fractal world, these connections as the curriculum evolves lend depth and pattern to the curriculum process. With the Sierpinsky triangle fractal, as the pattern emerges, we see these same patterns repeated at finer and finer scales. These emerging patterns of the curriculum are built through and emerge from the process. Thus, rather than considering the curriculum as building blocks that are put together to form a wall, the curriculum is seen to exhibit properties of self-similarity across scales, depth, and emergence as the different areas of the Sierpinsky triangle are related to each other and emerge through the recursive process. The Sierpinsky triangle is not "built" the way the wall is built, but emerges as a complex structure with a depth of inner relationships that characterizes it from other fractals.

Another aspect of a relational curriculum involves the interconnections among students and the evolution of meanings through the discursive process. Social relationships that emerge through communications in the curriculum process are meaning relationships that extend individual and shared meanings and understandings. Again, the curriculum, perceived as a meaning relationship, emerges, evolves, and can be transformed through the language games of the curriculum.

Curriculum relations therefore can be understood from pedagogical and meaning perspectives (Doll, 1993). Thus, in addition to the curriculum structure that emerges during the educative process perceived as the complex of pedagogic techniques and classroom interaction patterns, the relational meaning structures created are the curriculum perceived from both individual and social perspectives. Together, the "pedagogic" relations and "meaning" relations merge as the curriculum itself takes on its own autopoietic, socio-autonomous, identity relations.

The fourth characteristic of Doll's (1993) emergent, process-oriented curriculum is *rigor*. Rigor, as one aspect of the rhythm of education according to Doll (1993), "keeps a transformative curriculum from falling into either "rampant relativism" or "sentimental solipsism" (p. 181). As he explains, "rigor means purposely looking for different alternatives, relations, connections" (p. 182). It also means examining assumptions, challenging traditions, and encouraging creative solutions. Like Dewey's inquiry process, shared meanings and understandings evolve and transform only when we challenge in some precise way our meaning structures through inquiry. One way to challenge our assumptions and explore alternative possibilities, according to Wittgenstein (1953) is to question and transform our language games. The language-games approach is intended for Wittgenstein, as Dewey's logic was intended by him, to extend, expand, and transform our meaning systems and ways of understanding.

As with Whitehead's (1929/1957) dance among romance, precision, and generalization, there is rhythm of interrelationship among the four aspects of Doll's (1993) curriculum. Rhythm itself, as Whitehead explains (1929/1957) involves "an interweaving of cycles, the whole process being dominated by a greater cycle of the same general character as its minor eddies" (p. 27). Each of these aspects of the curriculum dynamics, *richness*, *recursion*, *relations*, and *rigor*, are supported by a curriculum with the potential for self-organization, socioautonomy, and self-similarity across scales. These are characteristics of a learning organization. Enabling schools to become learning organizations will entail infusing the curriculum with these capabilities as well. The rhythm of the curriculum, like the beating of our own hearts, is integral to schools as learning organizations and will be discussed below.

Curriculum as process

The challenges of a process-oriented, emerging, self-organizing curriculum include accepting and envisioning how this approach can evolve from our existing frameworks, and challenging what we consider to be learning and knowing. These two are connected. From a process perspective, we must escape from our current view of the curriculum as a "thing." A postmodern curriculum cannot simply replace our modernist version, like changing clothes from one day to the next, but must emerge from it and go beyond it. We don't just stop thinking of the curriculum as a "thing" and start thinking of it as a process. To change our very ideas about the meaning, purpose, and value of the curriculum and its relationship to schooling, we must change how we talk about the curriculum, including changing our metaphors and ways of seeing the curriculum. Other challenges facing us include re-examining the role of teachers, principals, and administrators in schools.

Ultimately, infusing schools with meaning and purpose, understanding schools as learning organizations, and operating from a postmodern logics perspective will require seeing things differently. These perspectives will also help us to transform seeing the curriculum as a "thing" to seeing it as a process. To do so means rejecting, at its most fundamental level, the underlying logic of domination. The logics of process, meaning, and systems offer the basis for a relational curriculum with self-creative, autonomous, and self-identity potentials. Such a curriculum becomes a meaning system with the potential to transform individuals as well as society.

The importance of the curriculum as a communication or meaning system has been overlooked for its potential for transforming schools. Often, literature on school change starts with changing administrative structures, power relations, and mission statements. Although not unconnected to the curriculum by any means, these approaches often fail to notice the idea that perhaps changing what we expect students to learn and how we expect them to learn may in fact challenge the other structures. The curriculum may be the key to changing schools.

Curriculum as a Learning Organization

A learning organization, as defined by Senge (1990) is in constant renewal and creativity, "continually expanding its capacity to create its future" (p. 14). As social structures and meaning systems, learning organizations create by deriving new meanings and coming to new understandings. Through these abilities to evolve meaning structures, learning organizations adapt and affect the changing environments in which they live. Communication networks become the medium through which meanings are shared. They also comprise the social fabric of the learning organization.

Schools are social structures with purpose and value. In order to adapt and create as learning organizations, they must also develop as communication systems. The curriculum serves as both a communication medium for schools and a learning organization in its own right. Similarly, the curriculum supports individual student's attempts to derive meaning and come to shared understandings. The curriculum is thus a medium for the learning organization of schools. It is also a learning organization itself with meaning productions, autonomy, and self-identity through the evolutionary process. Finally, the curriculum supports the efforts of individuals to learn, find meaning, adapt, grow, and transform their lives and environments.

Traditionally schools have been organized hierarchically. School reform has often started with changing leadership styles and offering leadership training for administrators and teachers. The curriculum has been given, assumed, and valued as an end in itself. While debates throughout the 20th century have swung back and forth between traditional content-focused and progressive child-focused curriculum, the assumptions and treatment of the curriculum as a "thing" rather than as a process have predominated.

The traditional language games of schooling, treating the curriculum as a "thing" and debating the merits of a child-centered versus content focused curriculum, will never be resolved until the language games are changed. Although the relationship among the child, curriculum, and society has been debated for more than 100 years, until we stop thinking of each component as a "thing," we will just bounce around the triangle like the ball in Pong bounces around the edges of the playing screen.

The curriculum matrix provides a different way of looking at the curriculum as self-organizing, emergent, adaptive, and meaningful. The emergence of meaning connects children and society with and through the curriculum process. The metaphors used to describe key aspects of the curriculum (*richness*, *relations*, *recursion*, and *rigor*) include the Mandelbrot set and fractals. Earlier ways of connecting the logics of postmodernism as intertwined and interconnected meanings in a hologram similarly reflect the complexity of the relationship between schools, the curriculum, children, and society. The fluidity of their relationship, rather than being layered or hierarchical, suggests the meanings they share lend coherence and clarity to the picture as a whole. The curriculum, therefore, like the information of a hologram, is distributed throughout the image. The curriculum is, in a very fundamental way, the hologram of schooling, implicating and intertwining process, learning, shared meanings, and social discourse. The curriculum is the heart of schooling. It gives the school life and meaning, and sustains and transforms it. The curriculum likewise allows individuals to evolve and grow within the meaning structures of society. Finally, the curriculum becomes an important medium in which larger social meaning structures and the fabric of society itself are intertwined.

How do we get beyond thinking of schools as hierarchies and tap their adaptive potential? How do we enact a vision of classrooms as operating at the borders of chaos and order, on the edges of the Mandelbrot set? How does the curriculum matrix evolve from our existing myopic view of standards, learning objectives, and fear of losing our ability to compete internationally? Changing the language games of schooling and thereby evolving our meaning structures of the curriculum, requires nurturing borderland classrooms.

Borderland Classrooms

I often use the book *Flatland* (Abbott, 1884/1994) in my undergraduate mathematics methods courses to promote discussion on dimensionality, perspective, and "thinking outside the box." Written in the 1880s, the book reveals Victorian values, including sexist ideas about the place of women in society. The value of the book, however, is the perspective it offers the acceptance of new ideas, seeing a different world, and the use of analogical reasoning.

In the society of Flatland, everyone lives on a two-dimensional plane; this is all the occupants of Flatland know. They have to develop special ways of seeing and negotiating their way around their flat world. The male inhabitants of Flatland are geometric shapes. Except for the warriors, which are isosceles triangles, the men are regular polygons: squares, pentagons, hexagons, and so on.

The social hierarchy is determined by inheritance. The number of sides you have determines your position in society. The ruling class is comprised of the circles, or those regular polygons that have so many sides that they look like circles. Polygons that are 96-sided are higher up than 12-sided ones who are higher up than hexagons, pentagons, or squares. Every generation, the male children will have one more side unless the father in some way disgraces the family. A square will have pentagons for sons. They, in turn, will have hexagons, and so on. (Women are straight lines. If angle measures are a measure of intelligence, women, as straight lines, have none.)

The story is about a square. One night, he has a dream, or, rather, a figure appears to him. The figure is a three-dimensional sphere. The only thing the square can see is the part of the sphere that appears in the plane. At first the sphere appears as a circle, then as smaller and smaller circles as it moves through the plane, until finally it looks like a single dot then disappears as it leaves the plane. The sphere tries to explain to the square how moving out of the plane, moving "up" out of his middle, there is another dimension, a third-dimension.

The square is incredulous. How can this be? Finally, in order to convince him, the sphere lifts the square out of the plane into the third-dimension. The square at first feels he has gone mad. He sees the inside of his house as he looks down on it, feeling nauseated. He finally understands what the sphere meant as he tried to explain how the square, in the plane, pulled out through itself, becomes a cube. He understands how line segments with two end points become squares in the second dimension with four end points (called vertices) and how then in the third dimension, become cubes with eight end points. Similarly, by analogy, in one dimension, the line segment has no sides. In two dimensions, the square has one side, and in three dimensions, the cube has six sides.

If, by analogy, there seems to be a progression in the number of sides and end points from one to two to three dimensions, what would the fourth dimension look like, the square asked? This, the sphere assured him, was impossible. There was no fourth dimension.

The square returned to Flatland and told of his discoveries. Only his grandson understood and appreciated this new way of seeing, this information about dimensionality. The priests in power, however, were not so kind. They imprisoned the square, labeling him insane for his different way of seeing.

There are several morals to this simple story. Many who read it understand the persecution and isolation felt by the square as he was ostracized and eventually put in jail for his different understandings and vision. Many recognize the Victorian cast system and the role of women in such a society. Analogical reasoning is seen as a powerful way of making sense of and understanding things we may not be able to directly experience. Also of interest is that the wise sphere could not himself see out of his worldview to use the same kind of reasoning he encouraged in the square to develop ideas about the fourth dimension. (Is this the same as when we encourage our students to think for themselves yet when they begin to challenge us, we tell them to be quiet?)

I retell this story because so much of our life is spent in Flatland. Our metaphoric routines and unquestioning ways of negotiating our world often go unchallenged as we find comfort and security in predictability and certainty. Our own students, who may have visions of Sphereland, who desire to move along the borders of the metaphoric Mandelbrot set, are often themselves put down and punished for trying to see things differently, from another perspective. We ourselves may have experienced feelings of nausea as we look down at our own comfortable world through a different lens, from a different vantage point. The familiar looks unfamiliar and our entire way of organizing our world is called into question.

I've challenged my students to envision the transition from our three-dimensional world not to the fourth dimension but to a fractal dimension, a new Borderland. How do our ideas about relationship, meaning, and purpose change when we go from looking at dimension in the traditional sense as measurement to dimension as relationship as did Mandelbrot when he invented fractals? Although in our standard way of understanding dimension, we understand one dimension to be lines with breadth only, two dimensions to involve length and breadth, and three dimensions to include length, breadth, and height, how do we envision fractional dimensions of 2.59, for example?

As discussed in Chapter Five, fractals are found in nature as branching structures, embedded patterns, with self-similarity relationships across scales. Dimensionality is redefined in fractal geometry as the relationship between self-similarity across scales and the self-similarity ratio. Rather than measuring fractals the way we measure around a square or the area of a circle, for example, fractals give us an indication of relational characteristics. What would assessment that considered fractal growth rather than accumulated growth or average growth look like? How would we change our ideas about intelligence if, instead of a measure (logico–mathematical–verbal) reasoning, we considered dimensionality of thought? How will the curriculum change when conceived as a Mandelbrot walk along the borders rather than as a path up a mountain?

Most graphic representations are so many points plotted along the plane to represent the behavior of some system. A line, plotted on the Cartesian plane, represents linear motion, across time, progressing infinitely in both directions. Each point of the Mandelbrot set, however, captures the dynamics of an entire set: one of the Julia sets. The Mandelbrot set is especially interesting because it is the set that represents the *behaviors* of all Julia sets in the complex plane. The Mandelbrot set also serves as a visual and mathematical metaphor for the dance of chaos, or complexity at the edge of chaos. Figure 8.4 shows a different view of the Mandelbrot set. The infinite variability, beauty, and repeating patterns of the Mandelbrot set, even the way it is generated as a set of behaviors of infinite Julia sets, make it a powerful image for complexity.

How might this dance evolve in our classrooms? How might the emergence and reappearance of patterns change our ideas about predictability and learning? How do individual learning patterns interconnect and relate to the emerging patterns as a whole?

Figure 8.4. Another perspective of the Mandelbrot set.

Together, these images are an impetus for changing our ideas about our roles in the classroom and challenging our ideas about the curriculum. Borderland classrooms provoke and implicate a rich, relational, recursive, and rigorous curriculum. Enacting new language games, understanding the curriculum as a meaning system with a life of its own, we must then be willing to accept the implications and creations of our imaginings. We must accept the possibility, explore the consequences of our own creations, and accept the idea that we may be treated as "insane" or punished by those who can't see beyond their modernist worlds.

I've wrestled with the potential consequences of encouraging my students to engage chaos and complex adaptive systems metaphors to change their language games of schooling. My concern is not (as the sphere in Flatland) that my students will later return to challenge the very assumptions we have replaced with postmodern perspectives. My moral dilemma is that what I offer may confuse, frustrate, and perhaps endanger the professional and personal lives of my students when they challenge the standards-driven, test-oriented curriculum they are forced to "disseminate" in their own classrooms.

Nevertheless, borderland classrooms are the type of environments where curriculum dynamics will occur. As a discursive community evolves, the nature of schooling as a meaning system will change and our conversations and understandings will evolve beyond their modernist origins. It is in these borderland classrooms that new patterns will emerge, new problems will stimulate further growth, and new challenges to our very roles and identities will occur.

Notes

1 See Devaney's site for some outstanding instructional tools as well as visual means for exploring the Mandelbrot set. http://math.bu.edu/people/bob/

2 These fractal images can be seen at: http://www.student.utwente.nl/~schol/gallery/. They were created using *FractInt* by Eric Schol and are reprinted by permission. His personal homepage is http://www.student.utwente.nl/~schol/index.html and e–mail is eric@schol.nl. All rights reserved.

3 Again, Devaney's website has some interesting tutorials and information available for purchase both in print format and video format on the Mandelbrot bulbs.

4 Moving a disk every second, it would take 2^64–1 (that is, 2 times itself 64 times minus one) seconds for the earth to be destroyed. If I have done my mathematics correctly, this would be a little over 14,029,009,106,175 years.

CHAPTER NINE
Transformational Crossings

The particular use of "logic" throughout this book has been intended both as a bridge and as a challenge to our modernist ways of seeing the world. Aristotle distinguished between scientific reasoning and logical reasoning. His books on logic explore the form of language and the nature of argumentation. He begins his *Analytica Posteriora* with the statement "All instruction given or received by way of argument proceeds from pre-existent knowledge" (McKeon, 1973, p. 11). While logical form never produces new knowledge, scientific reasoning may. The form of reasoning, however, is defined and determined, for Aristotle.

Dewey's logic represented a significant break with Aristotelian logic as a form of argumentation. According to Burke (1994), "In Dewey's view, logical theory should address not just a formal study of linguistic syntax but it should be grounded in a theory of experience" (p. 136). Dewey's logic defines his process of inquiry as an approach to making sense of our experiences. For Dewey, however, logic as inquiry still attempts to discover something "real."

Wittgenstein's early attempts to understand the logical foundations of mathematics were rejected by him in his later works. For Wittgenstein:

> We want to say that there can't be any vagueness in logic. The idea now absorbs us, that the ideal "*must*" be found in reality. Meanwhile we do not as yet see *how* it occurs there, nor do we understand the nature of this "must." We think it must be in reality; for we think we already see it there.... Where does this idea come from? It is like a pair of glasses on our nose through which we see whatever we look at. It never occurs to us to take them off. (*PI*, 101–103)

Our desire to seek these underlying truths and structures is a very old pair of glasses we in the West have been wearing for thousands of years. While the glasses perhaps became bifocals during the Modern Era, we are only now testing our vision without those old spectacles. To understand our meanings, according to Wittgenstein, is a matter of looking at our use of language, by exploring the language games we play. To change our understandings, to put on a new pair of glasses, is to change the language games.

Postmodern logics of relationship, meaning, and systems employ the logic of taking off our glasses, changing our language games, and exploring new meanings. While our eyes are still trained to look for solid objects and objective truths, as our language games change, we may adjust to new sights and sounds.

Nietzsche describes the tightrope walker, hanging over the precipice, as we are hanging over the edge of the consequences of our own modernist assumptions and society. Although some would have us step away from the threshold of postmodernism, to climb down off the tightrope, and return to simpler (modern) times, we cannot erase the consequences of our own mechanistic and domineering past.

> Man is a rope, tied between beast and overman—a rope over an abyss. A dangerous across, a dangerous on-the-way, a dangerous looking-back, a dangerous shuddering and stopping. What is great in man is that he is a bridge and not an end: what can be loved in man is that he is an *overture* and a *going under*. I love those who do not know how to live, except by going under, for they are those who cross over. I love the great despisers because they are the great reverers and arrows of longing for the other shore. (Nietzsche, 1968, pp. 126–127)

We ARE the rope, we are the "dangerous across" as we make our way and create a world. Our way, the Western way, has been the way of the beast, rejecting and denying wholeness, relationship, creativity, meaning, purpose, and value. Our modernist world, threatening to collapse under its own sterility and isolation, is at a crossroads. There is danger in just stopping, ignoring the implications of our own creations, and there is not a going back. The only way over is to go down under.

Going down under, we challenge our most basic and cherished beliefs and values. Creativity and openness require approaching and embracing our challenges and difficulties, traveling to the edge, being pushed to our limits, and accepting the chaos with the faith that within the chaos is hidden order. And within the process there is growth and renewal as new ways of seeing as occur through our language-games approach.

Recreating heart, we must see schooling as a place where relationships and meaning rather than test scores and standards drive the curriculum. Generative metaphors, new ways of seeing as, will guide our efforts to form a new life, to create new meanings, to engage in alternative ways of interacting and relating with one another within the educational social context.

The metaphors I have used throughout this book have been deliberate. Standing at the threshold, we have one foot in our modernist past while we struggle to open the door and look beyond. Our eyes have not adjusted to the light while some of us still cling to our old glasses. We have not quite seen what is beyond the door, because our seeing as will in fact bring about what we can see on the other side. We are walking a tightrope, we are the tightrope, as we make our way from a dangerous past of oppression, hate, and domination. We are empowered to create our futures, but we must let go of our own end of the rope, of our own past securities and comforts.

The logics of postmodernism are still elusive. The orchestration of our futures, the concluding overture of modernism, may occur through the combined and connected ways of seeing relationship, meaning, and systems as our ways of talking about and thinking change through our language-games approach. Like the holograph, these are not three separate approaches or ways of seeing but, together, form a clearer image, a way of seeing that can create a new world, a new way of being.

Schools as learning organizations tap these visions and provide for integrated, meaningful, connected, holistic, relational understandings. The last chapter described the role of the curriculum in permeating and infusing schools with meaning and purpose. The curriculum is the heart of the social organization of schooling. This chapter will offer final visions for how we may begin this process, how we may come to see and create schools as learning organizations, and how the field of curriculum studies may be transformed.

Quantum Schools

In a recent *Scientific American* article (Astumian, 2001), the invention of molecule motors was discussed. Modeled after the energy exchange in cells, molecule motors may completely make obsolete our ideas about how machines can operate.

> The basic insight, loosely described as the Brownian ratchet principle, is that random noise can be put to good use. The trick is to rectify the noise, to filter out the randomness you do not want so that you are left with what you do want. This principle resembles the phenomenon known as stochastic synchronization, whereby increasing the noise in a communications channel can actually make it easier to transmit a signal. (Astumian, 2001, p. 58)

Current brain research suggests a similar phenomenon in the development of children's brains. The problem with a two year old isn't that there is a lack of neural connections but that there are too many. As a child's brain develops, it actually starts pairing down neural connections, making the networks or relationships in the brain more "fixed." A two-year-old tantrum may be the result of frustration associated with too many connections. Essential for brain development may be strategic filtering rather than an assembly line of building of relationships. Experience at the right times provides opportunities for selecting and making more firm the neural pathways that become fairly permanent throughout our lives. Increasing the noise (that is, increasing certain types of experiences) allows for this pairing and strengthening of pathways to occur without overly limiting the options. Like the ratcheting affect, each relationship or connection that gets used moves the developing brain one step further down the path of evolving into an adult brain. The mind of each individual is unique as the "noise" of experience varies, yet the commonalities and relationships among all of us as humans appear to be our order, our strange attractors from the chaos of our different pasts.

While the molecule motor operates at extremely small scales, the idea can be applied to schools. How would a quantum school, modeled after molecular motors or human brains, compare to a school modeled after mechanical principles? How does the idea of using noise, rather than suppressing it, make sense from the perspective of learning organizations? How does the noise become useful information?

We know what schools, designed after mechanical principles, look like. The factory metaphor of schooling treats students as the raw materials to be formed and produced for consumption by society. The teachers manage the process. Principals are the quality-control managers. When the consumers or the (school) board detect flaws in the product, school administrators demand increased performance, tighter control, technical training for the managers, or a change in the curriculum. The curriculum is the blueprint for production; that, by which, the products are measured and designed. The curriculum defines what the products are supposed to accomplish. The goal is to maximize production within estimated budgets. Larger factories are more efficient; departmentalization and specialization allow for technical expertise to develop among the managers. The quality-control measures allow certain degrees of latitude and even predict a certain proportion of failures. Success is measured not by individual accomplishments but by whether the average quality of the products falls within these bounds. Those that are lost are just wastage.

Now what of a quantum school? The principle of Brownian motion suggests that randomness, unpredictability, and noise can actually encourage the dynamics for individual and collective action. An environment rich in complexity encourages the reorganization and utilization of energy. Rather than treating irrelevancy, unpredictability, and uncertainty as things to be controlled, in the quantum universe we should be encouraging them. The creative potential of individuals and the perpetuation of the mechanism itself depend on diversity and variability, not predictability and control.

Add to the quantum mix meaning and purpose, and "noise" can now be understood as anomalies, uncertainties, and challenges to existing meaning structures. Learning and growth occur through accommodation as contradictions are resolved, language games evolve, and our way of seeing as creates new forms of life, new relationships and connections, and new meanings. Just as in the butterfly effect, noise and feedback that may have seemed too insignificant to have an impact on the behavior of the system as a whole may accumulate and synergistically combine to have effects beyond any imaginable predictions.

As we engage in new language games, for example, exploring generative metaphors of quantum physics, chaos and complexity, we may, however, be trying too hard. Like the child learning to whistle, the more we concentrate on puckering and blowing, the less likely whistling sound emanates from our mouths. Evolving schools as learning organizations requires a critique of our current language games and a rejection of the logic of domination. It also may require letting go both of our known pasts and of our desired, envisioned futures.

Letting Go

In his book *Going to pieces without falling apart*, psychiatrist Mark Epstein (1999) discusses his experiences as a therapist incorporating Buddhist principles into his practice.

> Afflicted, as we are, with a kind of psychological materialism, we are concerned primarily with beefing ourselves up. Self-development, self-esteem, self-confidence, self-expression, self-awareness, and self-control are our most sought after attributes. But Buddhism teaches us that happiness does not come from any kind of acquisitiveness, be it material or psychological. Happiness comes from letting go. In Buddhism, the impenetrable, separate, and individuated self is more of the problem than the solution. (p. xvi)

The kind of "psychological materialism" he addresses is like the perspective we have of schools. We think we need to clarify roles, relationships, and identities in our social fabric of society. We believe measurable quantities are indications of quality. Yet the answer may be, as Epstein found, not in our achieving new understandings but in our ceasing to look for them and letting go of the past.

Epstein delineates in his book the Buddhist processes of emptiness, surrender, meditation, connection, tolerance, relationship, passion, and relief for achieving happiness. As we explore the meaning of schooling, we too may find answers in our cessation of the search, our surrender to uncertainty, our experiencing and nurturing of relationships, our tolerance for ambiguity and diversity, and our passion for an ethical, caring way of interacting.

Rather than trying to discern the meaning of much postmodern art by looking at the figures represented in a painting, it often makes more sense to look at the negative space, the space supporting the figures rather than the figures themselves. What are the negative spaces of schools? How do we look at schools by not looking at them? What do we look at instead?

One of the empty spaces of schools is not what schools are doing but what they are not doing. We are not accomplishing equity of opportunities, shared meanings, and fulfilled lives through educative experiences in schools. While trying to understand and define the purpose of schooling reduces schooling to its products, looking at the empty spaces of schooling captures the dynamic and evolutionary perspective of schools as social institutions.

The interconnections and evolving relationships among social institutions implicates looking at the empty spaces of our social fabric. Considering the empty spaces means not finding or seeking easy answers or quick fixes but tracing out the complexity of interrelated and interconnected influences. It also means we can't solve the problems of schooling without also addressing pollution, poverty, war, pestilence, hate, discrimination, and so on. To find the answers to these problems requires letting go of our ideas that these problems are things or that they have simple solutions. They are networks of interrelated and interconnected events that create and support the empty spaces of schools and what we perceive as problems with schools are also reflections of these larger problems.

Letting go of our ideas of problems as things and looking into the empty spaces for interconnections and interrelationships, however, can be consuming and depressing. There is no place to start. We have to address all of the issues, all of the problems, together, all at once. There are no simple causes, no single dominoes to set in motion to solve our problems. To prevent the insanity and feelings of utter despair, in addition to looking at the empty spaces, we also need to empty ourselves, remove doubt, ignore our own powerlessness, and recognize the impact of seeing things differently. The goal is not to eliminate emptiness but to engage in and learn from it. Engaging emptiness, rather than trying to escape from it, is the going down Nietzsche describes in the quote earlier in this chapter.

The vision of emptiness that quantum physics can provide is that, in fact, even when we experience the empty spaces and vast distances of the cosmos, we are connected and related through complex fields of interconnected wholeness. Engaging emptiness is also a willingness to explore these interconnections and fields of relationship. Sometimes exploring these relationships requires developing a way of seeing. We need to learn to see, not more clearly, but with "soft eyes."

Epstein (1999) describes how horseback riders take in their surroundings with "soft eyes" rather than focused concentration. He describes how one of his patients was having trouble as she rode completing her hurdles. She was too focused on the hurdle as the end product of the jump. The more she messed up, the more anxious she became. One of her instructors, he describes, taught her to imagine after each jump an additional maneuver she would need to make. Rather than focusing on the jump, she started looking beyond it. In education, we don't appreciate looking with "soft eyes" but want a clear focused vision of our students' learning or classroom accomplishments. We will see more clearly, however, when we can look beyond these ends or products as measures of accomplishment and see them as processes enveloped in larger processes.

How do we let go, look at the empty spaces, and see with soft eyes? How will these strategies of emptiness, surrender, meditation, connection, tolerance, relationship, passion, and relief ultimately affect the lives of our kids? How do the logics of relationship, meaning, and systems interconnect and provide understandings for changing everything all at once?

Practicalities

At the end of the movie "Mindwalk" (based on the 1984 book by Capra, *The turning point*), the politician in the movie bemoans the fact that the day has come to an end. After a daylong discussion among his friend, a Romantic poet, a quantum physicist, and himself, he expresses the need for next steps. "Is this just the best discussion I have had in a long time" he worries, or will something come of it?

Although in the movie I identify more with the physicist and poet than the politician, as we come to the final chapter, I too feel obligated to consider the practicalities of this book. Although most scholars engaged in curriculum studies have found the need to retreat to the abstractions of theory rather than take on the entire Western logic of domination, I would like to believe my discussion of postmodern logics and the language games of schooling may have practical application in the schools. I would also like to consider the implications for educational research when our language games reflect the logics of relationship, meaning, and systems. The overwhelming feelings of despair at the thought of having to tackle everything at once actually become the impetus for action. Curriculum studies, as the heart of education (just as curriculum is the heart of schooling), can serve as a link and a passageway, a rope, and its crossing.

The future of curriculum studies

The chasm between curriculum theorists and other curriculum specialists in the field of curriculum studies has been widening over the last twenty years of the 20th century. The backlash in education that has resulted in the national testing movement, international comparison studies, charter schools, vouchers, and the disempowerment of educators has elevated the role of the curriculum evaluators and, to some extent, the curriculum specialists, curriculum developers, and curriculum designers while curriculum theorists have become more removed from the practicalities of schooling. As the public becomes more and more involved in education, however, even those in more traditional fields of curriculum studies have found their expertise less and less valued. Business, politicians, and the lay public have more of an influence on curricular decisions than do the professionals, researchers, teachers, and teacher educators who work directly in the schools.

The division among the fields of curriculum studies has been along philosophical and theoretical lines. Status-quo curricularists still have a place in the existing structure of schools. Methods textbooks for teachers to more efficiently and effectively teach their students occupy a large portion of the professional education literature. Most highly valued among these texts, however, are those written by classroom teachers that are experientially rather than research or theoretically based. Only those university professionals who either team with a classroom teacher or were themselves once classroom teachers find a market for their texts. Educational researchers interested in publishing curriculum techniques and suggestions cannot get published in the larger and most prestigious journals without first demonstrating that their ideas have been tested and found to work in "real" classrooms.

Even a bigger business than the business of producing and marketing methods textbooks, however, is the production, distribution, and sale of curriculum textbooks used in the schools. The wars between the biggest publishers have been played out as the larger companies buy out the smaller ones. Only three or four major textbook companies have survived the competition of the marketplace. Although each huge publisher produces a variety of different textbook series in the different content areas, they are in business for profit, not quality education. A typical mathematics textbook now has over 500 pages. Rather than going into depth or taking a stand to represent a particular theoretical framework, the large publishing houses, driven by the need for profit, offer a grab bag approach to curriculum. Put everything in there and hope there is something for everyone. At the same time, equity and social issues are treated as problems of probability, not as driving forces of the curriculum. These companies have panels that make sure there are a sufficient proportion of Hispanics, Blacks, Asians, and women represented in the pictures and names used in the books. Stories that include multicultural perspectives, often told from a majority culture perspective, must be included. Excluded, however, are any messages of value, meaning, or purpose. It is not politically correct to take a stand on questions that pertain to values.

So even the traditional curriculum specialists find they must take a back seat to the demands of the powerful public. Their work is undervalued as political agendas and "commonsense" ideas about learning and teaching, usually going back to a vision of the 1950s, drive decision-making and the production of curriculum products. The future of curriculum studies in all areas may look bleak if we expect their impact to be felt in our modernist schools.

The role of curriculum studies from all perspectives, however, can take on a more promising future. We must challenge status-quo work. In letting go, in learning to see softly, in working to bring about a vision of quantum schools, we may approach our work from postmodern perspectives, bringing forth new school and social realities. One approach, the approach of this book, is to recreate the heart of schools from the holographic perspectives of meaning, relationship, and systems. As a way, and not THE way, of approaching curricular studies, our new curricular work is to engage in conversations about the underlying meaning structures upon which our queries are based. Only then can we come to shared meanings and understandings and, ultimately, only then can schools become the heart of our society as the curriculum is the heart of schooling. When we quit thinking of the curriculum as a thing, when children, teachers, parents, and society are perceived as complexes of relationships, and when social institutions are understood as meaning systems, the role of the curricularist takes on a special meaning. To "show the fly the way out of the fly-bottle," curriculum specialists are not in a position to line everyone up and lead the way. Their role is to offer a change of aspect, a way of seeing as.

Working with teachers in preservice and inservice teaching, I have marveled at the impact (or lack of impact) I have had on them and their practices. The teachers who embrace the ideas expressed in this book and perhaps change their teaching are not the ones who are simply looking for methods, tricks in the classroom, or what to do on rainy Friday afternoons. They are the ones who have a "change in aspect." They understand that to include fractal geometry in an 8th grade mathematics class, for example, is not to add something else to the already burdened curriculum, but that fractal geometry offers children, and themselves, a new way of looking at their world and their mathematics. Perhaps some of the "old" mathematics is not as important as we once thought, they consider. They further reason that chaos mathematics challenges our traditional notions of dimensionality, suggesting that mathematics and other domains of study in school may be more fluid, more changing, and less certain than traditional textbooks might have us think. They even study with renewed interest the history of the development of mathematics.

The social significance of changing meaning and communication structures in schools is that the impact is felt throughout our social fabric. Similarly, educational research must not be within our "normal science" of status-quo education but should offer, again, visions for how to escape the fly-bottle of modern education.

Educational research

As discussed earlier in this chapter, Aristotle was among the first to distinguish between logical form and empirical sciences. According to him, both were valid ways of knowing, and both led us to certain knowledge as long as the rules of reasoning were followed. Like mathematics, science and empirical studies are closed systems. No new knowledge is generated as we engage in scientific or mathematical inquiry. We operate from the givens, follow acceptable procedures, and derive objective and certain truths that are already stated within the conditions of inquiry. Because we only study things that can be measured, our investigations are further limited by our imagined ways of measuring. The measurement bias both limits our insights and focuses us on what we think is important.

We have challenged in this book the assumptions of meaning and knowing from their traditional Western perspectives with our postmodern logics. We now are faced with subjecting our own research and inquiry to this kind of scrutiny. One thing we must challenge is that what we are researching has any value beyond that which we give it. There is no underlying truth and all measurement is of our own invention. We also need to recognize that, in educational research, we have thus far focused on measuring "things." We have yet to develop postmodern tools for understanding relationships. As Heisenberg and Bohr discovered eighty years ago, what we choose to examine, and how we choose to measure it, will, in a very fundamental way, bring forth the reality we are looking for. We also must realize how our inquiries communicate and create value. Defining intelligence the way we have reduces our ideas about intelligence, places value on certain kinds of thinking over others, and limits our abilities to recognize and accommodate other kinds of intelligences including creativity, intuition, and insight.

These challenges to what we academicians hold as most sacred, our scholarship and research, offer a promise, however, an escape of sorts, from the domination of traditional science over the social sciences. While the participatory nature of inquiry implies that we have an impact in creating the kinds of schools we desire as we engage in educational explorations, the challenge to traditional modes of inquiry should offer opportunities for creativity and the search for new meanings.

For example, research on children's learning of fractions concepts not only assumes the importance of being able to manipulate fractions using paper-and-pencil algorithms, but also elevates the importance of learning fractions in the schools when we notice how much difficulty children have with them. Because measures of understanding fractions are traditionally restricted to computational facility and understanding of algorithmic approaches, we have become blind to other ways students experience, make sense of, and use part-whole relationships. Often, those who have good intuitive or practical understandings of fractions are disadvantaged by common school approaches to the study of fractions. What they learn is not more about fraction concepts but that either they are no good at mathematics or that mathematics is stupid or irrelevant to their lives.

Educational research questions that have an implied comparison and inherent reductionism to dualistic thinking are also popular, especially as education becomes more combative and battlelines more distinct. Although every doctoral student knows not to frame their questions as "yes-no" questions, the dualisms are hidden within the conceptual framework of the study. Also implied, even in many qualitative studies, are causality and generalizability, especially if any intervention is used, or hope for "improvement" drives the study. Studies of the effectiveness of methods or instructional strategies, for example, are generalized over populations and contexts. Implied in these studies is the ability to control conditions of schooling and overcome contextual and individual differences.

When the results of educational research are mixed (which they usually are), rather than interpreting this as a problem with our inquiry (either the methods or the very questions themselves), we interpret the results in whatever ways are politically and/or economically expedient. For every study that shows the advantages of whole language approaches to literacy development, for example, there are studies that tout phonics. If a particular legislation or school board decision needs support, the fishing expedition is to find the studies that support the cause. In some respects, this is no more reprehensible than claiming a single underlying truth or reality since, at its most fundamental levels, our research brings about our realities.

As seen in our investigations of scientific revolutions of the 20th century, simple causality, predictability, and underlying realities should be called into question. Sensitivity to initial conditions, complex networks of interactions, and personal autopoiesis and autonomy destroy our hope of an orderly, predictable world. Order and prediction are more uncharacteristic than a regularity of our complex world.

A colleague once challenged me: "Is all chaos methodology research postmodern research?" I followed this nurturing challenge with my own challenge about whether there was something fundamentally hegemonic about mathematics—not about mathematics instruction per se, but about modern mathematics itself (Fleener, 1999). These kinds of questions encourage us to consider the underlying logics of inquiry. Educational research for all too long has been satisfied to adopt first the scientific method model of inquiry and then to pattern investigations along the lines of psychological research. It is time for educational researchers to understand the social, political, economic, and ecological implications of our inquiries; the interconnectedness of social systems; and the potential for initiating new ways of interacting that are more holistic, relational, and meaningful. To engage in these kinds of investigations, however, requires a change in aspect, a different way of seeing, a willingness to cross over.

In a recent conference of the Society for Chaos Theory in Psychology and Life Sciences (SCTPLS),[1] researchers from economics, brain research, education, philosophy, sociology, psychology, physiology, and ecology met to share techniques, expertise, and studies using chaos and the complexity sciences. At thinktanks like the Santa Fe Institute, researchers from a variety of fields come together to explore how complex systems operate.[2] How does trying to study complexity change both our techniques and what we see? How do the conversations and shared expertise across these diverse disciplines extend our postmodern potentials by rejecting the fragmentation of modern science? How might what we choose to study, reflect and change what we value? These questions are only beginning to be answered.

We must realize, however, that the techniques of chaos and complexity sciences research, as techniques, can be reduced and proceduralized to the extent that all meaning and value become sterile. We also mustn't confuse the power of these tools for shaping how we see with the things themselves. Just as modernism confused the tools with the science and the metaphor of the clock became an assumed way of interpreting how nature really was, we must be careful, as we expand our techniques and change our ways of seeing, not to forget that we are in fact creating those visions and bringing forth those realities.

So why engage in educational research if we in some way discover that which we seek? This is a question I have wrestled with over the years. Occasionally I am surprised by the results of my investigations, but usually I am not. Is this because I am so good at what I do? Hardly! The questions we ask reveal what we think is important and reflect our ways of seeing. Sometimes, the most important questions we can ask are about the questions we are asking. As we explore our choice of studies as well as the techniques we employ, our ideas about what is important in schools and for students to learn may change. We may also discover that we are asking the wrong questions; that the questions we are asking reflect a perspective that is untenable and inconsistent with a way of life we envision beyond the modernist logic of domination. This is the way down under.

Going Down Under

In her book of empowerment, Starhawk (1988) ends her prologue with the following:

> For me, the journey that began in despair now reaches a place of hope and a sense of empowerment.... We can only begin.... If there is to be renewal, it begins with us. We can touch—through these words, these pages. We can know the dark, and dream it into a new image. (p. xxx)

I too feel we have "dreamed the dark" and through our shared discourse can create a new image of schooling. The logics of postmodernism reveal an "awareness of the world and everything in it as alive, dynamic, interdependent, interacting, and infused with moving energies: a living being, a weaving dance" (Starhawk, 1988, p. 9). The implications are not replacing one Grand Narrative with another but creating a new life, a new way of being, infused with meaning and purpose.

> I am not talking about a being somewhere outside of this world, nor am I proposing a new belief system. I am talking about choosing an attitude: choosing to take this living world, the people and creatures on it, as the ultimate meaning and purpose of life, to see the world, the earth, and our lives as sacred. (Starhawk, 1988, p. 11)

Part of awakening the dark is engaging in conversations with each other. I have deliberately pulled in writings from fields other than education because there is a movement afoot, so to speak, a synergy of our times that is propelling us beyond our modernist heritage and fragmented approaches to inquiry. At the same time, there are those profane spirits, those who would have us return to our past, who challenge the sacred (often in the name of religion), and destroy our confidence and faith in our own evolutionary powers.

> Structure, not content, determines how energy will flow, where it will be directed, what new forms and structures it will create. Hierarchical structures, no matter what principles they espouse, will breed new hierarchical structures that embody power over not power-from-within. A spiritual organization with a hierarchical structure can convey only the consciousness of estrangement, regardless of what teachings or deep inspirations are at its root. (Starhawk, 1988, p. 19)

The structure of a learning organization, complex, interrelated, evolving, and adaptive, better reflects our own needs and promises to be engaged in creative self-expression and purposeful activity. Understanding schools as learning organizations recognizes meaning, value, and purpose as the core of communications that constitute the learning organization. The curriculum, as the heart of schooling, is the meaning structure of schools.

Crossing Over

To let go of the rope while still standing on it, to continue to cross to our uncertain futures, to go down to the depths of despair and return with hope and promise are the processes of crossing over. To recreate heart, to understand the curriculum as the essence of schooling, and to explore the curriculum as a learning organization and meaning system, we may be able to gain some glimpses beyond the threshold on which we stand. Discarding the lenses of modernism and finding suitable replacement lenses may make all the difference in the world—as long as we realize that even our new lenses filter our way of seeing and living. We must endlessly recreate heart.

Notes

1 For information on the Society for Chaos Theory in Psychology and Life Sciences (SCTPLS), see the Web site: http://www.societyforchaostheory.org/.

2 Many papers and announcements of public seminars can be found on the Santa Fe Institute (SFI) Web site: http://www.santafe.edu/.

Bibliography

Abbott, E. A. (1884/1994). *Flatland: A romance of many dimensions* (5th Ed.). New York: Harper Collins.

Alexander, T. M. (1995). Educating the democratic heart: Pluralism, traditions and the humanities. In J. Garrison (Ed.), *The new scholarship on Dewey,* pp. 75-91. Norwell, MA: Kluwer Academic Publishers.

Angyal, A. (1941). *Foundations for a science of personality.* Cambridge, MA: Harvard University Press. (Excerpts from chapter 8 in F.E. Emery (1969), *Systems thinking,* Baltimore, MD: Penguin Books, (pp. 17-29)).

Arthur, W. B. (1990). Positive feedbacks in the economy. *Scientific American (February),* pp. 92-99.

Astumian, R. D. (2001). Making molecules into motors. *Scientific American, 285*(1) (July), pp. 56-64.

Banathy, Bela A. (1998) A taste of systemics. In *The primer project: The first international electronic seminar on wholeness* (December 1, 1996 to December 31, 1997). http://www.newciv.org/ISSS_Primer/asem04bb.html (accessed 04/18/02).

Bergson, H. (1911). *Creative evolution.* Henry Holt and Company.

Bertalanffy, L. (1968). *General systems theory: Foundations, development, applications.* New York: George Braziller Press.

Bertalanffy, L. (1975). The history and development of general systems theory. In E. Taschdjian (Ed.), *Perspectives on general systems theory: Scientific-philosophical studies,* pp. 149-169 (Essays compiled after Bertalanffy's death), New York: George Braziller.

Bethanis, S. J. (1995). Language as action: Linking metaphors with organization transformation. In S. Chawla & J. Renesch (Eds.), *Learning organizations: Developing cultures for tomorrow's workplace*, pp. 185-196. Portland, Oregon: Productivity Press.

Bohm, D. (1980). *Wholeness and the implicate order.* New York: Routledge.

Boulding, K. (1956). General systems theory - the skeleton of science. *Management Science*, 2(3), pp. 197-208.

Briggs, J. P. & Peat, F. D. (1984). *Looking glass universe.* New York: Simon & Schuster.

Burke, T. (1994). *Dewey's new logic: A reply to Russell.* Chicago: The University of Chicago Press.

Capra, F. (1984). *The turning point: Science, society, and the rising culture.* New York: Bantam Books.

Capra, F. (1996). *The web of life: A new scientific understanding of living systems.* New York: Anchor Doubleday Books.

Chardin, Teilhard de (1965/1957). *The divine milieu.* New York: Harper & Row.

Cobb, J. B. Jr. & Schroeder, W. W. (Eds.) (1981). *Process philosophy and social thought.* Chicago, IL: Center for the Scientific Study of Religion.

Colilli, P. (1997). *The idea of a living spirit: Poetic logic as a contemporary theory.* Buffalo, NY: University of Toronto Press.

Crosby, A. W. (1997). *The measure of reality: Quantification and Western society, 1250-1600.* New York: Cambridge University Press.

Cummins, J. & Sayers, D. (1997). *Brave new schools: Challenging cultural illiteracy through global learning networks.* New York: St. Martin's Press.

Davies, P. (1995). *About time: Einstein's unfinished revolution.* New York: Simon & Schuster.

Dell, P. F. (1985). Understanding Bateson and Maturana: Toward a biological foundation for the social sciences. *Journal of Marital and Family Therapy, 11*(1), (January), 1-20.

Dennett, D. C. (1995). *Darwin's dangerous idea: Evolution and the meanings of life.* New York: Touchstone Books.

DeVries, R. (1997). Piaget's social theory. *Educational Researcher, 26*(1), 4-17.

Dewey, J. (1910). *The influence of Darwin on philosophy.* New York: Henry Holt and Company.

Dewey, J. (1925/1981). *Experience and Nature.* Chicago: Open Court. Reprinted as *Later Works 1.*

Dewey, J. (1929). *The quest for certainty: A study of the relation of knowledge and action.* New York: Minton, Balch & Company.

Dewey, J. (1931). The development of American pragmatism. In J. Dewey (Ed.), *Philosophy and civilization*, 13-35. New York: G. P. Putnam's Sons.

Dewey, J. (1934). *Art as experience.* New York: Perigee Books.

Dewey, J. (1938). *Logic: The theory of inquiry.* New York: Holt and Company.

Doll, W. (1993). *A post-modern perspective on curriculum.* New York: Teachers College Press.

Doll, W. (2002). Ghosts and the curriculum. In W. Doll & N. Geogh (Eds.), *Curriculum Visions.* Peter Lang Press.

Einstein, A. (1956/1993). *Out of my later years: The scientist, philosopher and man portrayed through his own words.* Avenel, NJ: Random House Publishing.

Ekeland, I. (1990). *Mathematics and the unexpected.* (translation by the author; original published in 1984). Chicago: The University of Chicago Press.

Elkind, D. (September 1995). School and family in the postmodern world. *The Phi Delta Kappan, 77*(1), 8-14.

Epstein, M. (1999). *Going to pieces without falling apart: A Buddhist perspective on wholeness.* New York: Broadway Books.

Evans, M. D. (1998). *Whitehead and philosophy of education: The seamless coat of learning.* Amsterdam, Netherlands: Rodopi B. V.

Fesmire, S. A. (1995). Educating the moral artist: Dramatic rehearsal in moral education. In J. Garrison, (Ed.), *The new scholarship on Dewey*, pp. 45-60. Norwell, MA: Kluwer Academic Publishers.

Fleener, M. J. (1999). Toward a poststructural mathematics curriculum: Expanding discursive possibilities. *Journal of Curriculum Theorizing* (Summer), pp. 89-105.

Fleener, M. J. & Fry, P. G. (1998). Adaptive Teacher Beliefs for Continued Professional Growth During Postmodern Transitions. In D. J. McIntyre & D.M. Byrd (Eds.), *Strategies for career-long teacher education: Teacher education yearbook VI*, pp. 154-167, Thousands Oaks, CA: Corwin Press.

Fleener, M. J. & Rodgers, D. B. (1999). A systems theoretic approach to understanding transformation in learning communities. *Journal of Thought, 34*(1), pp. 9-22.

Fleischaker, G. (1990). Origins of life: An operational definition. *Origins of Life and Evolution of the Biosphere, 20*, pp. 127-137.

Fletcher, S. (2000). *Education and emancipation: Theory and practice in a new constellation*. New York: Teachers College Press.

Foucault, M. (1977/1995). *Discipline and punishment: The birth of the prison*. New York: Vintage Books.

Fraser, J. T. (1975). *Of time, passion and knowledge: Reflections on the strategy of existence*. Princeton, N.J.: Princeton, N.J. Press.

Freire, P. (1970). *Pedagogy of the oppressed*. New York: Seabury Press.

Fritzsch, H. (1994). *An equation that changed the world: Newton, Einstein, and the theory of relativity*. Chicago: University of Chicago Press.

Fullan, M. (1993). *Change forces: Probing the depths of educational reform*. London: Falmer Press.

Genova, J. (1995). *Wittgenstein: A way of seeing*. New York: Routledge Press.

Gleick, J. (1987). *Chaos: Making a new science*. New York: Penguin Books.

Gragg, A. (1976). *Makers of the modern theological mind: Charles Hartshorne*. Waco, TX: Word Books, Publisher.

Gribbon, J. (1995). *Schrodinger's kittens and the search for reality: Solving the quantum mysteries*. New York: Little Brown and Company.

Griffin, S. (1980). *Woman and nature: The roaring inside her*. New York: Harper Colophon Books.

Hagberg, G. L. (1994). *Meaning and interpretation: Wittgenstein, Henry James, and literacy knowledge*. Ithaca, NY: Cornell Press.

Harding, S. (1986). *The science question in feminism*. Ithaca, New York: Cornell University Press.

Hargreaves, A. (1994). *Changing teachers, changing times: Teacher's work and culture in the postmodern age*. New York: Teachers College Press.

Holder, J. J. (1995). An epistemological foundation for thinking: A Deweyan approach. In J. Garrison, (Ed.), *The new scholarship on Dewey*, pp. 7-24. Norwell, MA: Kluwer Academic Publishers.

Husserl, E. (1997/1970). *The Crisis of European Sciences and Transcendental Phenomenology* (translated with an introduction by David Carr; first published in German in 1954, edited by Walter Biemel.). Evanston, IL: Northwestern University Press.

Kaufmann, W. (1974). *The portable Nietzsche* (39th printing). New York: Viking Press.

Kliebard, H. M. (1998). The effort to reconstruct the modern American curriculum. In L. E. Beyer & M. W. Apple (Eds.), *The curriculum: Problems, politics, and possibilities* (2nd Edition). Albany: State University of New York Press, pp. 21-33.

Kliebard, H. M. (1975). Metaphorical roots of curriculum design. In W. Pinar (Ed.), *Curriculum theorizing: The reconceptualists*, pp. 84-85. Berkeley: CA: McCutchan.

Kline, M. (1982). *Mathematics: The loss of certainty.* New York: Oxford University Press.

Knappman, E. W. (1994). *American trials of the 20th century.* New York: A New England Publishing Associates Book.

Kofman, F. & Senge, P. (1995). Communities of commitment: The heart of learning organizations. In S. Chawla & J. Renesch (Eds.), *Learning organizations: Developing cultures for tomorrow's workplace,* pp. 15-44. Portland, Oregon: Productivity Press.

Kuhn, T. S. (1957). *The Copernican Revolution.* Cambridge, MA: Harvard University Press.

Kuhn, T. S. (1962). *The structure of scientific revolutions.* Chicago: University of Chicago Press.

Lakoff, G. & Johnson, M. (1980). *Metaphors we live by.* Chicago: The University of Chicago Press.

Lathem, E. C. (1969*). The poetry of Robert Frost.* New York: Holt, Rinehart, and Winston.

Latour, B. (1993). *We have never been modern* (trans. By Catherine Porter). Cambridge, MA.: Harvard University Press.

Levins, R. & Lewontin, R. (1985). *The dialectical biologist.* Cambridge, MA: Harvard University Press.

Loder, J. E. & Neidhardt, W. J. (1992). *The knight's move: The relational logic of the spirit in theology and science.* Colorado Springs: Helmers & Howard.

Lucas, G. R. (1979). *Two views of freedom in process thought.* Missoula, MT: Scholars Press.

Lucas, G. R. (1983*). The genesis of modern process thought: A historical outline with bibliography.* Metuchen, NJ: The Scarecrow Press, Inc. and the American Theological Library Association.

Luhmann, N. (1986). The autopoiesis of social systems. In F. Geyner & J. van der Zouwen (Eds.), *Sociocybernetic paradoxes*. London: Sage Publications.

Luhmann, N. (1990*). Essays on self-reference.* New York: Columbia University Press.

Luhmann, N. (1995). *Social systems* (transl. J. Bednarz). Stanford, CA: Stanford University Press.

Mandelbrot, B. (1983). *The fractal geometry of nature.* New York: W.H. Freeman.

Marshall, J. D., Sears, J. T., & Schubert, W. H. (2000). *Turning points in curriculum: A contemporary American memoir.* Upper Saddle River, NJ: Prentice-Hall, Inc.

Maturana, H. R. (1980). Introduction and biology of cognition. In H. R. Maturana & Varela, F. J., *Autopoiesis and cognition: The realization of the living*, pp. xi-xxx and 5-58. Boston, MA: D. Reidel Publishing Company.

Maturana, H. R. (1988). Reality: The search for objectivity or the quest for a compelling argument. *The Irish Journal of Psychology*, *9*(1), 25-82.

Maturana, H. R. & Varela, F. J. (1980*). Autopoiesis and cognition: The realization of the living.* Boston, MA: D. Reidel Publishing Company.

Maturana, H. R. & Varela, F. J. (1987). *The tree of knowledge: The biological roots of human understanding*. Boston, MA: Shambhala Publications, Inc.

McKeon, R. (1973). *Introduction to Aristotle (Second edition)*. Chicago: University of Chicago Press.

Mead, G. H. (1932). *The philosophy of the present*. Chicago: Open Court Publishing.

Mead, G. H. (1934/1967). *Mind, self and society from the standpoint of a social behaviorist*. (Ed. with an introduction by C. W. Morris). Chicago: University of Chicago Press.

Mead, G. H. (1964). What social objects must psychology presuppose? In A.J. Reck (Ed.), *Selected writings: George Herbert Mead*, pp. 105-113. Chicago: University of Chicago Press.

Merchant, C. (1980). *The death of nature: Women, ecology and the scientific revolution*. San Francisco: Harper Collins Publishers.

Mingers, J. (1995). *Self-producing systems: Implications and applications of autopoiesis*. New York: Plenum Press.

Monk, R. (1990). *Ludwig Wittgenstein: The duty of genius*. New York: Macmillan.

Morris, C. W. (1967). (Ed.) *Works of George Herbert Mead, Volume 1: Mind self and society from the standpoint of a social behaviorist*. Chicago: University of Chicago Press.

Myrdal, G. (1962). *An American dilemma: The negro problem and modern democracy*. New York: Harper & Row (original edition, 1944).

Nietzsche, F. (1968). Thus spoke Zarathustra. In W. Kaufmann, *The portable Nietzsche* (39th printing), pp. 103-439. New York: Viking Press.

Oliver, D. W. & Gershman, K.W. (1989). *Education, modernity, and fractured meaning: Toward a process theory of teaching and learning*. Albany, NY: State Univesity of New York Press.

Peat, F. D. (1987). *Synchronicity: The bridge between matter and mind*. New York: Bantam Books.

Penfield, W. (1975). *The mystery of the mind: A critical study of consciousness and the human brain*. Princeton, NJ: Princeton University Press.

Piaget, J. (1965). *The moral judgment of the child*. New York: Free Press.

Pinar, W. F., Reynolds, W. M., Slattery, P., & Taubman, P. M. (1995). *Understanding curriculum: An introduction to the study of historical and contemporary curriculum discourses*. New York: Peter Lang.

Posner, G. F. (1998). Models of curriculum planning. In L. E. Beyer & M. W. Apple (Eds.), *The Curriculum: Problems, Politics, and Possibilities* (2nd Edition), pp. 79-100. Albany: State University of New York Press.

Postman, N. (1995). *The end of education: Redefining the value of school*. New York: Vintage Books.

Prigogine, I. & Stengers, I. (1984). *Order out of chaos: Man's new dialogue with nature*. New York: Bantam Books.

Quine, W. V. O. (1953). *From a logical point of view*. New York: Harper Torchbooks.

Reck, A. J. (1975). Process philosophy, a categorical analysis. *Tulane Studies in Philosophy*, 25, pp. 58-91.

Rennie, J. (1999). From the editor. *Revolutions in science*. New York: Scientific American.

Rifkin, J. & Perlas, N. (1983). *Algeny*. New York: Viking.

Ring, K. (1984). *Heading toward omega: In search of the meaning of the near-death experience*. New York: W. Morrow.

Rorty, R. (1967). Metaphilosophical difficulties of linguistic philosophy. In R. Rorty (Ed.), *The linguistic turn: Recent essays in philosophical method*, pp. 1-39. Chicago: University of Chicago Press.

Schubert, W. H. (1986). *Curriculum: Perspective, paradigm, and possibility*. New York: Macmillan.

Schwandt, T. A. (2001). *Dictionary of qualitative inquiry*. Thousand Oaks, CA: Sage Publications.

Senge, P. (1990). *The fifth discipline: The art and practice of the learning organization*. New York: Doubleday.

Sirotnik, K. A. (1998). What goes on in classrooms? Is this the way we want it? In L. E. Beyer & M. W. Apple (Eds.), *The curriculum: Problems, politics, and possibilities* (2nd Edition), pp. 58-76. Albany: State University of New York Press.

Smith, J. E. (1959). John Dewey: Philosopher of experience. In C. W. Hendel (Ed.), *John Dewey and the experimental spirit in philosophy. (Four lectures delivered at Yale University commemorating the 100th anniversary of the birth of John Dewey)*. New York: Liberal Arts Press.

Smith, T. V. & Grene, J. (Eds.). (1963). *Berkeley, Hume, and Kant*. Chicago: The University of Chicago Press.

Stacey, R. D. (1992). *Managing the unknowable: Strategic boundaries between order and chaos in organizations*. San Francisco: Jossey-Bass Publishers.

Starhawk. (1988). *Dreaming the dark*. Boston: Beacon Press.

Stewart, I. (1989). *Does God play dice?: The mathematics of chaos*. Cambridge, MA: Blackwell Press.

Toulmin, S. (1992). *Cosmopolis: The hidden agenda of modernity*. Chicago: The University of Chicago Press.

Toulmin, S. (1990). *Cosmopolis*. Chicago: The University of Chicago Press.

Varela, F. (1981). Describing the logic of the living: The adequacy and limitations of the idea of autopoiesis. In M. Zeleny, (Ed.), *Autopoiesis: A theory of the living organization*, pp. 36-48. New York: Elsevier-North Holland.

Von Glasersfeld, E. (1995). *Radical constructivism: A way of knowing and learning*. Washington, DC: Falmer Press.

Waldrop, M. (1993). *Complexity: The emerging science at the edge of order and chaos*. New York: Simon & Schuster.

Walkerdine, V. (1988). *The mastery of reason: Cognitive development and the production of rationality*. New York: Routledge Press.

Warren, K. J. (1996). The power and promise of ecological feminism. In K. J. Warren (Ed.) *Ecological feminist philosophies*. Bloomington, IN: Indiana University Press, pp. 19-41.

Warren, K. J. (1997). Taking empirical data seriously: An ecofeminist philosophical perspective. In K. J. Warren (Ed.), *Ecofeminism: Women, culture, nature*. Bloomington, IN: Indiana University Press, pp. 3-20.

Weedon, C. (1997). *Feminist practice and poststructural theory (Second Edition)*. Cambridge, MA: Blackwell Publishers Inc.

Wheatley, M. J. (1994). *Leadership and the new science: Learning about organization from an orderly universe*. San Francisco: Berrett-Koehler Publishers, Inc.

Whitehead, A. N. (1925/1967). *Science and the Modern World, Lowell Lectures, 1925.* New York: The Free Press.

Whitehead, A. N. (1929/1957). *The aims of education.* New York: The Free Press.

Whitehead, A. N. (1929/1978). *Process and reality: An essay in cosmology.* New York: The Free Press.

Wittgenstein, L. (1953). *Philosophical investigations (PI). (*trans. G. E. M. Anscombe). New York: MacMillan.

Wittgenstein, L. (1959). *The blue and brown books.* Oxford: Blackwell Press.

Wittgenstein, L. (1961). *Tractatus logico-philosophicus (TLP)* (trans. D. F. Pears & B. F. McGuinness). London: Routledge.

Wittgenstein, L. (1967). *Zettel* (Z), (trans. G. E. M. Anscombe). Oxford: Blackwell Press.

Wittgenstein, L. (1969). *On certainty (OC),* (trans. Denis Paul & G. E. M. Anscombe). Oxford: Blackwell Press.

Wittgenstein, L. (1978). *Remarks on the foundations of mathematics (RFM),* (trans. G. E. M. Anscombe). Oxford: Blackwell Press.

Wittgenstein, L. (1980). *Culture and value (CV)* (trans. Peter Winch). Oxford: Blackwell Press.

Wolf, F. A. (1987). The physics of dream consciousness: Is the lucid dream a parallel universe? *Second Lucid Dreaming Symposium Proceedings/Lucidity Letter 6, no. 2.*

Zohar, D. (1990). *The quantum self: Human nature and consciousness defined by the New Physics.* New York: Quill/William Morrow.

Index

COUNTERPOINTS

Studies in the Postmodern Theory of Education

General Editors
Joe L. Kincheloe & Shirley R. Steinberg

Counterpoints publishes the most compelling and imaginative books being written in education today. Grounded on the theoretical advances in criticalism, feminism, and postmodernism in the last two decades of the twentieth century, Counterpoints engages the meaning of these innovations in various forms of educational expression. Committed to the proposition that theoretical literature should be accessible to a variety of audiences, the series insists that its authors avoid esoteric and jargonistic languages that transform educational scholarship into an elite discourse for the initiated. Scholarly work matters only to the degree it affects consciousness and practice at multiple sites. Counterpoints' editorial policy is based on these principles and the ability of scholars to break new ground, to open new conversations, to go where educators have never gone before.

For additional information about this series or for the submission of manuscripts, please contact:

Joe L. Kincheloe & Shirley R. Steinberg
c/o Peter Lang Publishing, Inc.
275 Seventh Avenue, 28th floor
New York, New York 10001

To order other books in this series, please contact our Customer Service Department:

Or browse online by series: